PLASMA PHYSICS IN ACTIVE WAVE IONOSPHERE INTERACTION

PLASMA PHYSICS IN ACTIVE WAVE IONOSPHERE INTERACTION

Spencer P Kuo

New York University-Tandon School of Engineering, USA

World Scientific

NEW JERSEY • LONDON • SINGAPORE • BEIJING • SHANGHAI • HONG KONG • TAIPEI • CHENNAI • TOKYO

Published by

World Scientific Publishing Co. Pte. Ltd.

5 Toh Tuck Link, Singapore 596224

USA office: 27 Warren Street, Suite 401-402, Hackensack, NJ 07601

UK office: 57 Shelton Street, Covent Garden, London WC2H 9HE

British Library Cataloguing-in-Publication Data
A catalogue record for this book is available from the British Library.

PLASMA PHYSICS IN ACTIVE WAVE-IONOSPHERE INTERACTION

ISBN 978-981-3232-12-9

For any available supplementary material, please visit
http://www.worldscientific.com/worldscibooks/10.1142/10767#t=suppl

Typeset by Stallion Press
Email: enquiries@stallionpress.com

Printed in Singapore

to my parents;
and to Sophia Ping Kuo,
my wife, constant companion, and best friend.

Preface

One essential feature of the plasma media is supporting various plasma waves. This textbook was prepared to provide students with an understanding of plasma waves, which will enable them to expand their studies into related areas. This is illustrated by introducing plasma research through ionospheric heating experiments, which I have conducted over the past several decades and found that an understanding of plasma waves is key to preparing for and performing theoretical and experimental plasma research and understanding the experimental results. Plasma waves excited in high-frequency (HF) wave–ionospheric plasma interaction stimulates numerous linear and nonlinear phenomena, which help to advance plasma physics. The selection of topics and the focus given to each provide a way for a lecturer to cover the bases and applications in a plasma wave course.

I have been teaching undergraduate seniors and first-year graduate students in plasma engineering over the past two decades. The material presented in the text is a compilation of those lectures as well as some of the research results associated with the ionospheric HF heating experiments.

Plasma is a dispersive dielectric medium with quite different properties from those of the conventional dielectrics. When a magnetic field is imposed, this anisotropic medium can sustain plasma waves of various mode types and polarizations. The first 3 Chapters provide the basis of plasma modes, where the formulations and analyses of the dispersion equation are presented and the characterizations of plasma modes are discussed. The inhomogeneity of the plasma complicates the propagation of plasma waves, and its nonlinearity gives rise to mode–mode coupling. Techniques to study wave propagation and formulation of mode coupling are presented in Chapter 4.

In the laboratory, the study of plasma waves is complicated by the boundary of the plasma contained in a device. Ionosphere is a natural unbounded plasma. Ionospheric heating and modification by powerful HF waves transmitted from the ground has been a very active research area. It is a platform for experimental and theoretical investigation of wave–wave and wave–particle interactions in magneto plasma. In Chapter 5, several salient features of the experimental observations are presented to illustrate the role of plasma waves excited via parametric instabilities in stimulating nonlinear plasma processes. Chapters 6 and 7 are devoted to the formulation and analysis of parametric instabilities. The excited plasma waves may evolve into nonlinear periodic or solitary wave; the nonlinear Schrödinger equation and Korteweg–de Vries (KdV) equation governing the nonlinear evolution of high frequency electron waves and low-frequency ion waves are derived and analyzed in Chapter 8. A potential application of the research in ionospheric heating experiments is to set up an ionospheric very-low-frequency transmitter for underwater communications. This topic is discussed in Chapter 9. The presentation in this book is self-contained and should be read without difficulty by those who have adequate preparation in classic mechanics and electromagnetism. The book will be useful to undergraduate seniors, graduate students, and researchers in plasma physics and engineering, as well as those in geophysics.

I wish to express my sincere gratitude to Professor Bernard R. S. Cheo who guided me into this field and has given me much helpful advice and kind encouragement on my scientific evolution throughout my career. I would also like to acknowledge my close collaborators, Professor Min-Chang Lee and Dr. Arnold Lee Snyder for their helpful discussions which influenced the presentation of several chapters.

Spencer Szu-Ping Kuo

List of Figures

Contents

Chapter 1

Basis of Plasma

1.1 Introduction

Plasma is a distinct state of matter; it contains electrically charged and neutral particles, including electrons, ions, and atoms/molecules. Charged particles enable plasma to conduct electricity and to react collectively to electromagnetic forces. Such unique properties separate plasma from the solid, liquid, and gas states, and so it is referred to be the fourth state of matter. Plasmas are estimated to constitute more than 99% of the visible universe.

Different from the other states of matter, plasma temperatures and densities span a range from relatively cool and tenuous (like aurora) to very hot and dense (like the central core of a star). Plasma can be accelerated and steered by electric and magnetic fields which allow it to be controlled for applications. One of the essential applications is plasma for energy through controlled thermal nuclear fusion. The core of our Sun is an example of fusion reactor, which produces the thermal energy of the Sun. Besides its black body radiation delivering heat to the earth, our Sun constantly emits plasma, which moves out in all directions at very high speeds to fill the entire solar system and beyond. It supplies a dilute but persistent stream of plasma containing protons, electrons, and other ions, named solar wind, to impinge on earth's magnetosphere and interact with the geomagnetic field. Some of electrons may flow down in sheets, through the earth's magnetic field loss cones appearing at polar latitudes in both the northern and southern hemispheres, into the earth's upper atmosphere. These electrons excite molecules, which radiate to form beautiful aurora in the upper atmosphere.

Other plasma applications include plasma engines for long-distance interplanetary space travel missions, magnetohydrodynamic (MHD) generators that increase the efficiency of electric generation and reduce the pollutions of burning coal; plasma etching in fabricating integrated circuits; surface treatment to promote adhesion; medical treatment for coagulation and sterilization; to name a few.

Ionospheric plasma is a platform for experimental and theoretical investigation of wave–wave and wave–particle interactions. Powerful HF waves transmitted from the ground heat the ionosphere to stimulate linear and nonlinear plasma processes. The observations facilitate the advance of nonlinear plasma theory.

The HF heating facility is also applied to explore new technological innovations, such as implementing an ionospheric very-low-frequency (VLF) transmitter for underwater communications and setting up an *in situ* RF source, which transmits whistler waves to scatter MeV electrons in the magnetosphere and to scatter tens of keV protons in the aurora ring current, for mitigating space weather effects on the lifetime of satellites and on the GPS communications.

Very energetic electrons have a strong impact on passing satellite systems. Satellites are designed to survive a certain amount of radiation (ionizing) dose accumulated during their lifetimes. Unexpected enhancement of the radiation fluxes caused by, for example, very strong solar storms, will significantly increase the total radiation dose to the satellites. Consequently, the radiation damage on active electronics and detectors of satellite systems will accumulate much faster than designed for. As the damage exceeds a threshold level, satellite systems become incapable of performing their mission. Strong storms also enhance the energetic proton level in aurora ring current. Energetic protons generate density irregularities which cause the scintillation of GPS signals.

1.1.1 *Collective phenomena*

1.1.1.1 *Debye shielding*

Consider a positive point charge q_0 located at the origin ($r = 0$); in vacuum, it produces a potential field

$$\varphi_0(r) = \frac{q_0}{4\pi\varepsilon_0 r} \tag{1.1}$$

As this charge is embedded in plasma of density n_0, this potential field (1.1) is changed. This is because the electric field of this charge modifies the spatial distribution of the surrounding charged particles, which forms a cloud of oppositely charged particles (assumed to be electrons in this case). This density perturbation $\delta Q_e = -e\delta n_e$ induces a potential field adding to the original potential field (1.1).

This is illustrated by embedding this test charge in an electron gas (neutralized by immobile ions of the same density) with density n_0 and temperature T_e (in energy unit, i.e., T stands for $k_B T$, where $k_B = 1.38 \times 10^{-23}$ J/K is the Boltzmann constant, which appears in general together with temperature T in the form of $k_B T$; therefore, k_B will be removed throughout the book to simplify the presentation), the total (self-consistent) potential field $\varphi(r)$ is governed by the Poisson equation

$$\nabla^2\varphi(r) = -\frac{q_0}{\varepsilon_0}\delta(r) + \frac{e\delta n_e}{\varepsilon_0} \tag{1.2}$$

where $\delta(r)$ is the three-dimensional delta function; $-e = -1.6 \times 10^{-19}$ C denotes a unit electric charge carried by an electron; the electron density perturbation δn_e is calculated by assuming Maxwell–Boltzmann statistics for the electron distribution (ions are too heavy to be mobile); thus

$$\delta n_e = n_0 \exp\left[\frac{e\varphi(r)}{T_e}\right] - n_0 \cong n_0\frac{e\varphi(r)}{T_e}$$

and (1.2) becomes

$$\nabla^2\varphi(\mathrm{r}) - \frac{1}{\lambda_{\mathrm{De}}^2}\varphi(\mathrm{r}) = -\frac{\mathrm{q}_0}{\varepsilon_0}\delta(\mathrm{r}) \tag{1.3}$$

where $\lambda_{\mathrm{De}} = (\varepsilon_0 \mathrm{T_e}/\mathrm{n_0 e^2})^{1/2} = \mathrm{v_{te}}/\omega_{\mathrm{pe}}$ is called "(electron) Debye length"; $\mathrm{v_{te}} = (\mathrm{T_e}/\mathrm{m_e})^{1/2}$, $\omega_{\mathrm{pe}} = (\mathrm{n_0 e^2}/\mathrm{m_e}\varepsilon_0)^{1/2}$, and $\mathrm{m_e}$ are the electron thermal speed, electron plasma frequency, and electron mass, respectively.

Because of spherical symmetry, φ depends only on the radial coordinate r. It makes easy to solve (1.3) in spherical coordinates, where $\nabla^2 = \mathrm{r}^{-2}(\mathrm{d/dr})(\mathrm{r^2 d/dr})$ and (1.3) becomes

$$\frac{1}{\mathrm{r}^2}\frac{\mathrm{d}\left[\mathrm{r}^2\frac{\mathrm{d}\varphi(\mathrm{r})}{\mathrm{dr}}\right]}{\mathrm{dr}} - \frac{1}{\lambda_{\mathrm{De}}^2}\varphi(\mathrm{r}) = -\frac{\mathrm{q}_0}{\varepsilon_0}\delta(\mathrm{r}) \tag{1.4}$$

Setting $\varphi(\mathrm{r}) = \phi(\mathrm{r})\exp(-\mathrm{r}/\lambda_{\mathrm{De}})$ into (1.4) yields

$$\frac{1}{\mathrm{r}^2}\frac{\mathrm{d}\left[\mathrm{r}^2\frac{\mathrm{d}\phi(\mathrm{r})}{\mathrm{dr}}\right]}{\mathrm{dr}} - \frac{2}{\lambda_{\mathrm{De}}}\frac{1}{\mathrm{r}}\frac{\mathrm{d}[\mathrm{r}\phi(\mathrm{r})]}{\mathrm{dr}} = -\frac{\mathrm{q}_0}{\varepsilon_0}\delta(\mathrm{r}) \tag{1.5}$$

If $\mathrm{d}[\mathrm{r}\varphi_0(\mathrm{r})]/\mathrm{dr} = 0$, where $\varphi_0(\mathrm{r})$ is a solution of this equation: $\mathrm{r}^{-2}\mathrm{d}[\mathrm{r}^2\mathrm{d}\phi(\mathrm{r})/\mathrm{dr}]/\mathrm{dr} = -(\mathrm{q}_0/\varepsilon_0)\delta(\mathrm{r})$, then $\varphi_0(\mathrm{r})$ is also a solution of (1.5). Thus, $\phi(\mathrm{r}) = \varphi_0(\mathrm{r}) = \mathrm{q}_0/4\pi\varepsilon_0\mathrm{r}$, and the solution of (1.3) is obtained to be

$$\varphi(\mathrm{r}) = \frac{\mathrm{q}_0}{4\pi\varepsilon_0\mathrm{r}}\exp\left[-\frac{\mathrm{r}}{\lambda_{\mathrm{De}}}\right] \tag{1.6}$$

As shown, the potential field of a test charge in plasma is contained in a sphere of radius λ_{De}, ascribed to the collective response of the plasma. However, the collective response of the plasma does not always work on shielding the perturbation; in fact, it can reinforce the perturbation under proper condition as will be elaborated by the next example.

1.1.1.2 *Plasma oscillation*

Consider a local perturbation of plasma which displaces a slab of electron gas from the equilibrium location of immobile ions by a

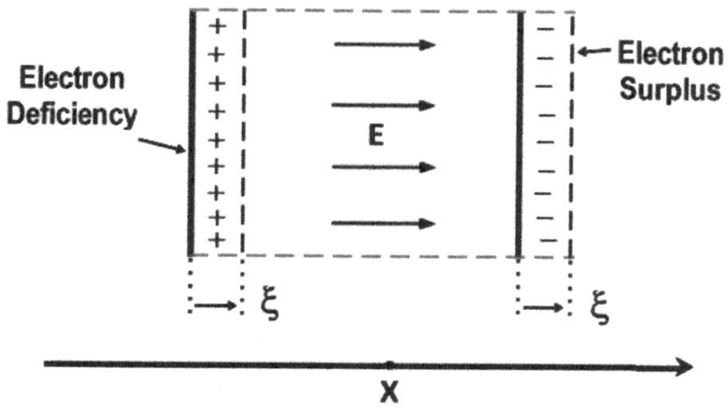

Fig. 1.1. Plasma oscillation.

small distance, ξ, as shown in Fig. 1.1. It induces an electric field $E = n_0 e\xi/\varepsilon_0$. This force field tends to pull electrons back to the equilibrium position, where the electron displacement ξ is governed by the equation of motion given to be

$$\frac{d^2\xi}{dt^2} = -\frac{eE}{m_e} = -\frac{n_0 e^2 \xi}{m_e \varepsilon_0} = -\omega_{pe}^2 \xi \tag{1.7}$$

It shows that this electron slab oscillates at electron plasma frequency $f_{pe} = \omega_{pe}/2\pi$. Plasma collectively sustains the perturbation to oscillate at a specific radian frequency ω_{pe}. Now, we extend the perturbation to have a spatial distribution, i.e., $\xi = \xi(x, t)$, and the electron plasma has a temperature T_e. Because the perturbation causes the electron plasma density n_e to vary spatially, i.e., $n_e = n_0 + \delta n(x, t)$ where $\delta n = -(\varepsilon_0/e)\mathbf{\nabla} \cdot \mathbf{E}$, there is an additional pressure gradient force, $-\nabla P_e = -3T_e \nabla \delta n$, acts on the electrons. Thus, (1.7) is modified to be

$$\frac{d^2\xi}{dt^2} = -\frac{eE}{m_e} - \frac{3T_e}{m_e n_0}\frac{d\delta n}{dx} = -\omega_{pe}^2 \xi + 3v_{te}^2 \frac{d^2\xi}{dx^2} \tag{1.8}$$

Equation (1.8) is a wave equation which has solutions of the forms $A_\pm \cos(\omega t \pm kx) + B_\pm \sin(\omega t \pm kx)$, where $\omega = (\omega_{pe}^2 + 3k^2 v_{te}^2)^{1/2}$ and $k = 2\pi/\lambda$ is the wavenumber with the wavelength λ. It shows that the perturbation can self-sustain and propagate in plasma when its oscillation frequency ω and propagation constant k satisfy a dispersion relation $\omega = (\omega_{pe}^2 + 3k^2 v_{te}^2)^{1/2}$ ascribed to the collective response of plasma.

1.2 Description of Plasma

Plasma is a many particle system in which charge particles interact with one another through the electric force; the trajectory of a charge particle is directed by the electric and magnetic fields imposed externally as well as those induced internally. Depending on the extent of exposing the plasma properties, there three common approaches employed to investigate plasmas: (1) single particle description, (2) kinetic description, and (3) fluid description.

1.2.1 *Single particle description (for the discreteness phenomena)*

Strictly speaking, this is a valid description of particle motion only inside the Debye sphere. The single particle equations of motion are given by

$$\frac{d\mathbf{r}}{dt} = \mathbf{v} \tag{1.9a}$$

and

$$m_a \frac{d\mathbf{v}}{dt} = \mathbf{F} = q_a(\mathbf{E} + \mathbf{v} \times \mathbf{B}) \tag{1.9b}$$

where m_a and q_a are the mass and charge of a particle of species "a".

1.2.1.1 *Free-particle motion in uniform electrostatic and magnetostatic fields*

We first consider a simple arrangement that plasma is embedded in uniform static electric field $\mathbf{E}_0$ and static magnetic field $\mathbf{B}_0$. Using the direction of $\mathbf{B}_0$ as a reference (i.e., $\|$) direction and decomposing

$\mathbf{E}_0$ into $\mathbf{E}_0 = \mathbf{E}_{0\|} + \mathbf{E}_{0\perp}$, (1.9) breaks down into

$$m_a \frac{d\mathbf{v}_\|}{dt} = q_a \mathbf{E}_{0\|} \tag{1.10}$$

and

$$m_a \frac{d\mathbf{v}_\perp}{dt} = q_a(\mathbf{E}_{0\perp} + \mathbf{v}_\perp \times \mathbf{B}_0) \tag{1.11}$$

The motion described by (1.10) is a uniformly accelerated translation in the $\mathbf{E}_{0\|}(\mathbf{B}_0)$ direction, i.e., $\mathbf{v}_\| = \mathbf{v}_{\|0} + q_a\mathbf{E}_{0\|}t/m_a$.

Substitute $\mathbf{v}_\perp = \mathbf{v}_{\perp 0} + \mathbf{v}_E$, where $\mathbf{v}_E = \mathbf{E}_{0\perp} \times \mathbf{B}_0/B_0^2$, into (1.11), and it reduces to

$$m_a \frac{d\mathbf{v}_{\perp 0}}{dt} = q_a \mathbf{v}_{\perp 0} \times B_0 \tag{1.12}$$

Thus, $\mathbf{v}_{\perp 0}$ has a constant magnitude, i.e., $|\mathbf{v}_{\perp 0}| = v_{\perp 0} =$ the initial value; it is the tangential velocity of a circular motion on a normal plane of $\mathbf{B}_0$, which gyros around $\mathbf{B}_0$ at a cyclotron angular frequency $|\Omega_a| = |q_a B_0/m_a|$ and has a (Larmour) radius $R = \mathbf{v}_{\perp 0}/|\Omega_a|$. Let $\mathbf{B}_0$ in the z direction, i.e., $\mathbf{B}_0 = \hat{\mathbf{z}}B_0$, and the initial velocity $\mathbf{v}_{\perp 0}(0) = \mathbf{v}_{\perp 0}(\hat{\mathbf{x}}\cos\theta + \hat{\mathbf{y}}\sin\theta)$, we then have

$$\mathbf{v}_{\perp 0}(t) = \mathbf{v}_{\perp 0}[\hat{\mathbf{x}}\cos(\Omega_a t - \theta) - \hat{\mathbf{y}}\sin(\Omega_a t - \theta)]$$

Since Ω_a is negative for electrons and positive for positive ions, the solution shows that electrons gyrate about the magnetic field based on right-hand rule and ions based on left-hand rule. The total velocity of the motion in the transverse direction (i.e., on the normal plane of $\mathbf{B}_0$), $\mathbf{v}_\perp = \mathbf{v}_{\perp 0} + \mathbf{v}_E$, manifests a combination of a gyration motion $\mathbf{v}_{\perp 0}$ and a uniform drift $\mathbf{v}_E$. This is a spiral motion which is illustrated in Fig. 1.2 for both positively and negatively charged particles. It is noted that $\mathbf{v}_E = \mathbf{E}_{0\perp} \times \mathbf{B}_0/B_0^2$ is independent of the charge and mass of the charge particles; thus, electrons and ions drift together in this collisionless case in the direction perpendicular to both the electrostatic and magnetostatic fields and there is no net electric current is produced.

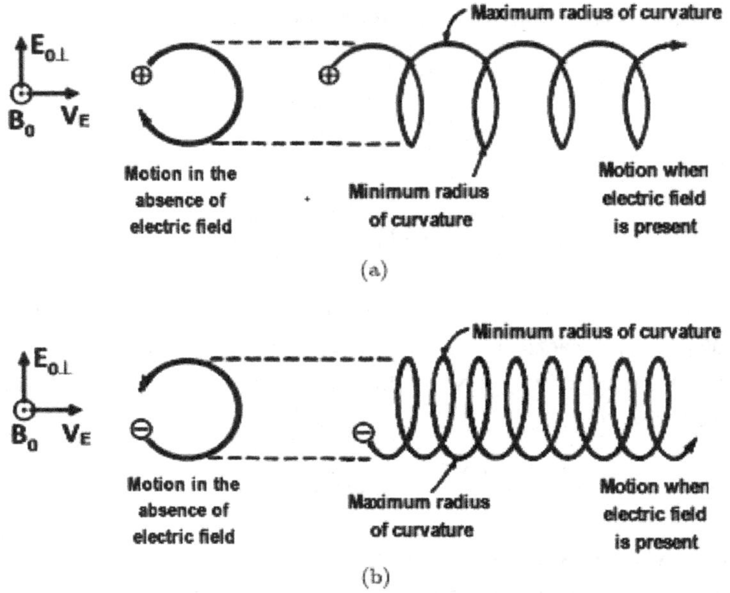

Fig. 1.2. Qualitative evolution of the trajectory in a normal plane of B_0: (a) positive charged particle, (b) negative charged particle.

1.2.1.2 *Free-particle motion in circularly polarized time harmonic electric field together with uniform magnetostatic field*

In (1.9), $\mathbf{B} = \hat{\mathbf{z}}\, B_0$, and the electric field is given by

$$E = \hat{\mathbf{z}}\, E_{0\|} \cos \omega t + E_{0\perp}(\hat{\mathbf{x}} \cos \omega t \pm \hat{\mathbf{y}} \sin \omega t)$$

where ± represent right-hand and left-hand circular polarization. Thus, (1.9) breaks down into

$$m_a \frac{dv_\|}{dt} = q_a E_{0\|} \cos \omega t \tag{1.13}$$

and

$$m_a \frac{dv_\perp}{dt} = q_a[E_{0\perp}(\hat{\mathbf{x}} \cos \omega t \pm \hat{\mathbf{y}} \sin \omega t) + \mathbf{v}_\perp \times \hat{\mathbf{z}} B_0] \tag{1.14}$$

The motion described by (1.13) is a uniform motion together with an oscillation in the magnetic field (z) direction, i.e.,

$$\mathbf{v}_{\|} = \mathbf{v}_{\|0} + \left(\frac{q_a E_{0\|}}{m_a \omega}\right) \sin \omega t$$

where the amplitude $|q_a \mathbf{E}_{0\|}/m_a \omega|$ is called "quiver velocity (speed)".

Substitute $\mathbf{v}_\perp = \mathbf{v}'_\perp + \mathbf{v}_1(\hat{\mathbf{x}} \sin \omega t \mp \hat{\mathbf{y}} \cos \omega t)$ into (1.14), it reduces (1.14) to

$$\frac{dv'_\perp}{dt} = q_a(\mathbf{v}'_\perp \times \hat{\mathbf{z}}\, B_0) \tag{1.15}$$

and the transverse quiver speed $v_1 = q_a E_{0\perp}/m_a(\omega \pm \Omega_a)$ is obtained.

Equation (1.15) is the same as (1.12); thus, $\mathbf{v}'_\perp = \mathbf{v}_{\perp 0}(t)$, represents inherent gyration around the magnetic field at the cyclotron frequency $f_{ca} = |\Omega_a|/2\pi$. The electric field rotates the charge particle at the oscillating frequency $f = \omega/2\pi$ of the electric field and at the quiver speed $v_1 = q_a E_{0\perp}/m_a(\omega \pm \Omega_a)$; this rotation around the magnetostatic field is perpendicular to the electric field.

When $\omega \pm \Omega_a \cong 0$, v_1 becomes very large. This is due to the cyclotron resonance effect, and the plus sign for the electron and minus sign for the ion. Thus, the electron resonates with right-hand circularly polarized field and the ion with left-hand circularly polarized field. At cyclotron resonance, (1.14) is solved formally as follows.

Let, $\mathbf{v}_\pm = v_x \pm i v_y$; the two scalar components of (1.14) are combined into

$$\frac{dv_\pm}{dt} \pm i\Omega_a v_\pm = \left(\frac{q_a E_{0\perp}}{m_a}\right) e^{i\omega t}$$

$$= e^{\mp i\Omega_a t} \frac{d(v_\pm e^{\pm i\Omega_a t})}{dt} \tag{1.16a}$$

At cyclotron resonance, $\omega \pm \Omega_a = 0$, i.e., $\pm\Omega_a = -\omega$, (1.16a) reduces to $d(v_\pm e^{-i\omega t})/dt = q_a E_{0\perp}/m_a$, which is integrated to $v_\pm = v_{\pm 0} e^{\mp i\Omega_a t} + (q_a E_{0\perp} t/m_a) e^{i\omega t}$, where $v_{\pm 0} = v_\pm(0) = v_{\perp 0} e^{\pm i\theta}$. Thus,

$$\mathbf{v}_\perp(t) = \mathbf{v}_{\perp 0}(t) + (q_a E_{0\perp} t/m_a)(\hat{\mathbf{x}} \cos \omega t \pm \hat{\mathbf{y}} \sin \omega t) \tag{1.16b}$$

It shows that the amplitude of the driven term increases linearly in time. This term is also in phase with the driving field.

Although this approach can still reveal the collective part of the motion, i.e., the drift motion $\mathbf{v}_E$, it cannot explore the collective behavior of plasma oscillations; other two approaches are adopted in general in the study of plasma waves.

1.2.2 *Kinetic description (for discreteness as well as collective phenomena)*

The (motion) state of a particle is characterized by its position $\mathbf{r}$ and velocity $\mathbf{v}$, which represents a point $(\mathbf{r}, \mathbf{v})$ in a six-dimensional phase space. A many-particles system, such as plasma, is characterized by the states of particles distributed in the phase space. Thus, plasma can be described in terms of density distribution functions $f_a(\mathbf{r}, \mathbf{v}, t)$ (one for each species) in the phase space, where $f_a(\mathbf{r}, \mathbf{v}, t)$ is defined as the number of particles in a unit spatial volume around $\mathbf{r}$ and in a unit velocity volume around $\mathbf{v}$ concurrently at time t, i.e., the density of a species at point $(\mathbf{r}, \mathbf{v})$ and at time t. In a small phase volume $\delta\mathcal{V} = \Delta\mathbf{r}\Delta\mathbf{v}$, the number of particles of a species is $f_a\delta\mathcal{V}$. The displacement of these particles from t to $t' = t + \Delta t$ in a small time interval Δt is illustrated in Fig. 1.3.

1.2.2.1 *Boltzmann (Vlasov) equation*

In the collisionless situation, the number of particles $f_a\delta\mathcal{V}$ in this small phase volume is invariant along the trajectory: $d\mathbf{r}/dt = \mathbf{v}$ and

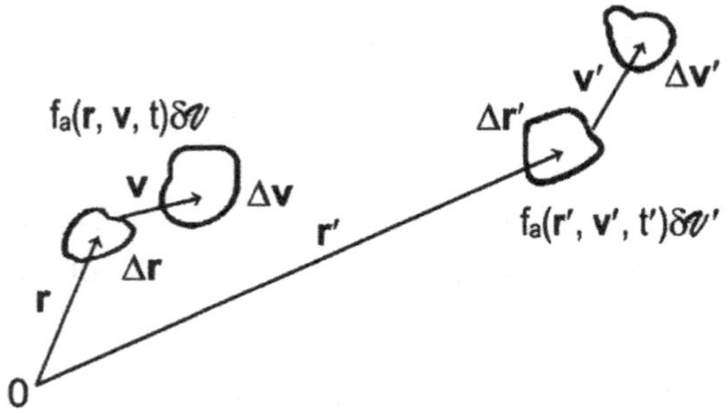

Fig. 1.3. The evolution of the states of particles in a differential phase volume $\delta\mathcal{V} = \Delta r\Delta v$ from t to $t' = t + \Delta t$.

$d\mathbf{v}/dt = \mathbf{F}/m_a$; thus $d(f_a\delta\mathcal{V})/dt = 0$. Because the phase volume is incompressible, i.e., $d\delta\mathcal{V}/dt = 0$, the density distribution function f_a is governed by the Boltzmann equation

$$\frac{df_a}{dt} = \frac{\partial f_a}{\partial t} + \mathbf{v}\cdot\nabla f_a + \left(\frac{\mathbf{F}_a}{m_a}\right)\cdot\nabla_v f_a = 0 \tag{1.17}$$

Boltzmann equation represents a continuity equation or a charge particle conservation equation, where $\mathbf{F}_a$ is a general force field. In plasma, $\mathbf{F}_a = q_a[\mathbf{E} + \mathbf{v}\times(\mathbf{B} + \mathbf{B}_0)]$ is a sum of electric and Lorentz forces, where $\mathbf{B}_0$ is the background magnetostatic field and $\mathbf{E}$ and $\mathbf{B}$ are the perturbation fields; in this case, (1.17) is also called Vlasov equation.

1.2.2.2 *Maxwellian distribution*

In non-relativistic situation, $|\mathbf{v}\times\mathbf{B}| \ll |\mathbf{E}|$; thus, $\mathbf{F}_a = q_a(\mathbf{E} + \mathbf{v}\times\mathbf{B}_0)$ is assumed in the following analyses. The common procedure to analyze (1.17) is to linearize (1.17) with respect to the first-order perturbation force $\delta\mathbf{F}_a = q_a\mathbf{E}$ by setting $f_a = f_{a0} + f_{a1}$, where f_{a0} is the unperturbed background distribution and f_{a1} is the first-order response to $\delta\mathbf{F}_a$. Substitute $f_a = f_{a0} + f_{a1}$ into (1.17), the zeroth- and first-order equations are obtained respectively to be

$$\frac{\partial f_{a0}}{\partial t} + \mathbf{v}\cdot\nabla f_{a0} + \left(\frac{q_a}{m_a}\right)(\mathbf{v}\times\mathbf{B}_0)\cdot\nabla_v f_{a0} = 0 \tag{1.18a}$$

and

$$\frac{\partial f_{a1}}{\partial t} + \mathbf{v}\cdot\nabla f_{a1} + \left(\frac{q_a}{m_a}\right)(\mathbf{v}\times\mathbf{B}_0)\cdot\nabla_v f_{a1} = -\left(\frac{q_a}{m_a}\right)\mathbf{E}\cdot\nabla_v f_{a0} \tag{1.18b}$$

The solution of (1.18a) is often a Maxwellian, i.e.,

$$f_{a0} = n_0\left(\frac{m_a}{2\pi T_a}\right)^{3/2}\exp\left(-\frac{m_a v^2}{2T_a}\right) \tag{1.19}$$

and (1.18b), a linearization of (1.17), is employed to study the linear wave properties (i.e., to derive the linear dispersion relations of plasma waves) of uniform magneto plasma.

This approach provides a sophisticated linear description of the plasma. The tradeoff is the complication of the analysis. Therefore, the fluid approach is introduced.

1.2.3 *Fluid description*

Because of the collective interaction, charged particles in plasma tend to move together as fluids. However, it requires that the number of electrons in a Debye sphere is much larger than 1, i.e., $n_0\lambda_{De}^3 \gg 1$. A parameter $g = 1/n_0\lambda_{De}^3$, called the plasma parameter, is used to define the fluid limit, which requires $g \leq 1$.

The fluid description preserves the collective behavior of the plasma, which is the most unique feature of plasma. The fluid equations are derived by taking velocity moments of the distribution function f_a via the Vlasov equation (1.17).

1.2.3.1 *Velocity moments*

The first three velocity moments of f_a are given as follows:

(1) 0th moment:

$$\int (\mathbf{v}^0) f_a(\mathbf{v}) d\mathbf{v} = n_a$$

(2) 1st moment:

$$\int (\mathbf{v}) f_a(\mathbf{v}) d\mathbf{v} = n_a \langle \mathbf{v} \rangle = n_a \mathbf{v_a}$$

(3) 2nd moment:

$$\int (\mathbf{v}\mathbf{v}) f_a(\mathbf{v}) d\mathbf{v} = n_a \mathbf{v}_a \mathbf{v}_a + \left(\frac{n_a T_a}{m_a} \right) \overleftrightarrow{\mathbf{I}}$$

where n_a, $\mathbf{v}_a$, and T_a are the fluid density, velocity, and temperature of a species "a"; $d\mathbf{v} = dv_x dv_y dv_z$; and $\overleftrightarrow{\mathbf{I}} = \hat{\mathbf{x}}\hat{\mathbf{x}} + \hat{\mathbf{y}}\hat{\mathbf{y}} + \hat{\mathbf{z}}\hat{\mathbf{z}}$ is a unit operator dyadic.

In the evaluation of each moment, we set $\mathbf{v} = \mathbf{v}_a + \delta\mathbf{v}$; here $\mathbf{v}_a = <\mathbf{v}>$ and $\delta\mathbf{v}$ is a random velocity, i.e., $<\delta\mathbf{v}> = 0$; the bracket $< >$ stands for an average over the velocity space. Thus, $\mathbf{vv} = \mathbf{v}_a\mathbf{v}_a + \mathbf{v}_a\boldsymbol{\delta}\mathbf{v} + \boldsymbol{\delta}\mathbf{v}\mathbf{v}_a + \boldsymbol{\delta}\mathbf{v}\boldsymbol{\delta}\mathbf{v}$, and $<\mathbf{vv}> = \mathbf{v}_a\mathbf{v}_a + <\delta\mathbf{v}\delta\mathbf{v}>$. Components δv_j of the random velocity $\delta\mathbf{v}$ are independent of each other, i.e., $<\delta v_i \delta v_j> = <\delta v_i><\delta v_j> = 0$ for $i \neq j$; thus, $<\delta\mathbf{v}\delta\mathbf{v}> = \hat{\mathbf{x}}\hat{\mathbf{x}}<\delta v_x^2> + \hat{\mathbf{y}}\hat{\mathbf{y}}<\delta v_y^2> + \hat{\mathbf{z}}\hat{\mathbf{z}}<\delta v_z^2>$. The random motion of particles gives rise to the temperature of the fluid; the relationship is given as $m_a<\delta v^2>/2 = 3T_a/2$. Because of the energy equipartition, $<\delta v_x^2> = <\delta v_y^2> = <\delta v_z^2> = <\delta v^2>/3 = T_a/m_a$, $<\boldsymbol{\delta}\mathbf{v}\boldsymbol{\delta}\mathbf{v}> = (T_a/m_a)\overleftrightarrow{\mathbf{I}}$ is obtained.

This procedure in general leads to a hierarchy of infinite number of moment equations, which are coupled to each other. In many practical applications, this infinite set is truncated at the second moment, which is aided by the ideal gas law, the pressure $P_a = n_a T_a$, to reach a closure.

1.2.3.2 *Fluid equations*

With the aid of the three defined moments and the ideal gas law, $P_a = n_a T_a$, a complete set of fluid equations derived from the Vlasov equation (1.17) includes

(1) Continuity equation [taking the 0th moment of (1.17)]

$$\frac{\partial n_a}{\partial t} + \boldsymbol{\nabla} \cdot n_a \mathbf{v}_a = 0 \tag{1.20}$$

(2) Momentum equation [taking the 1st moment of (1.17)]

$$n_a m_a \left(\frac{\partial \mathbf{v}_a}{\partial t} + \mathbf{v}_a \cdot \boldsymbol{\nabla} \mathbf{v}_a \right) = -\nabla P_a + q_a n_a (\mathbf{E} + \mathbf{v}_a \times \mathbf{B}_0) \tag{1.21}$$

where $\boldsymbol{\nabla} P_a = \gamma T_a \boldsymbol{\nabla} n_a$ and $\gamma = C_p/C_v$ is the ratio of specific heats C_p and C_v at constant pressure and volume, respectively. In the adiabatic compression, P/n^γ = Const., where $\gamma = (D + 2)/D$, and D is the number of dimensions of the compression, i.e., $\gamma = 3$, 2, and 5/3 for one-, two-, and three-Dimensional compression. In the isothermal case, $\gamma = 1$.

1.2.4 *Self-consistent description*

The electromagnetic fields in the kinetic equation (1.17) and in the fluid equation (1.21) are governed by Maxwell's equations

$$\nabla \times \mathbf{E} = -\frac{\partial \mathbf{B}}{\partial t} \tag{1.22a}$$

$$\nabla \times \mathbf{B} = \mu_0(\mathbf{J} + \mathbf{J}_e) + \frac{1}{c^2}\frac{\partial \mathbf{E}}{\partial t} \tag{1.22b}$$

$$\nabla \cdot \mathbf{E} = \frac{(\rho + \rho_e)}{\varepsilon_0} \tag{1.22c}$$

$$\nabla \cdot \mathbf{B} = 0 \tag{1.22d}$$

where $\mathbf{J}_e$ and ρ_e are externally imposed current density and charge density; $\mathbf{J} = e\int \mathbf{v}(f_i - f_e)d\mathbf{v} = e(n_i\mathbf{v}_i - n_e\mathbf{v}_e)$ and $\rho = e\int(f_i - f_e)d\mathbf{v} = e(n_i - n_e)$ are the induced current density and charge density by the electric field $\mathbf{E}$ in plasma; singly charged ions are assumed; ε_0 is the free space permittivity. These two physical quantities ρ and $\mathbf{J}$ are related through the continuity (conservation of charge) equation

$$\frac{\partial \rho}{\partial t} + \nabla \cdot \mathbf{J} = 0 \tag{1.23}$$

The first two curl equations (1.22a) and (1.22b) are associated with the Faraday's law and Ampere's Law, and the next two divergence equations (1.22c) and (1.22d) are Gauss's law for the electric charges and magnetic charges.

1.2.4.1 *Polarization charge and current*

Plasma is a dielectric medium and ρ in (1.22c) and (1.23) is the induced charge density in the dielectric. Thus, (1.22c) can be expressed in terms of the electric flux density $\mathbf{D}$ to be

$$\nabla \cdot \mathbf{D} = \rho_e \tag{1.24}$$

where $\mathbf{D} = \varepsilon_0\mathbf{E} + \mathcal{P}$; the induced polarization $\mathcal{P}$ by the electric field $\mathbf{E}$ in the dielectric is related to the induced charge density ρ and current density **J** in (1.22c) and (1.22b), respectively, to be

$$\nabla \cdot \mathcal{P} = -\rho \quad \text{and} \quad \frac{\partial \mathcal{P}}{\partial t} = \mathbf{J} \tag{1.25}$$

1.2.4.2 *Collision effect*

In collision plasma, collisions also affect physical processes; the momentum equation (1.21) is modified to be

$$n_a m_a \left(\frac{\partial \mathbf{v}_a}{\partial t} + \mathbf{v}_a \cdot \nabla \mathbf{v}_a \right) = -\nabla P_a + q_a n_a (\mathbf{E} + \mathbf{v}_a \times \mathbf{B}_0) - n_a m_a \nu_{ab} (\mathbf{v}_a - \mathbf{v}_b) - n_a m_a \nu_{an} \mathbf{v}_a \tag{1.26}$$

where ν_{ab} and ν_{an} are the collision frequencies of a charged particle of species "a" with the charged particles of species "b" and neutral particles, respectively; $n_a m_a \nu_{ab} = n_b m_b \nu_{ba}$ ascribed to the momentum conservation. Both elastic and inelastic collision events are added to count on the collision frequency ν_{an} and $\mathbf{v}_n \approx 0$ is assumed. Moreover, the ideal gas law for the electron fluid is replace by the electron thermal energy equation

$$\frac{\partial T_e}{\partial t} + \left(\frac{2T_e}{3} \right) \nabla \cdot \mathbf{v}_e + \delta(T_e) \nu_e(T_e)(T_e - T_n) + \text{ionization loss} = \left(\frac{2}{3n_e} \right) [Q + \nabla \cdot (\kappa_z \nabla_z + \kappa_\perp \nabla_\perp) T_e] + \text{heat input} \tag{1.27}$$

where $\delta(T_e)$ is the average relative energy fraction lost in each collision, e.g., in the elastic electron–neutral collisions, $\delta(T_e)\nu_e(T_e) = 2\nu_{en}(m_e/m_n)$; $\nu_e(T_e)$ is the effective collision frequency of electrons with neutral particles (it accounts for both elastic and inelastic collisions), T_n is the temperature of the background neutral particles; Q is the total Ohmic heating power density in the background plasma; $\kappa_z = 3n_e T_e / 2m_e \nu_e$ and $\kappa_\perp = (\nu_e/\Omega_e)^2 \kappa_z$ are the parallel and transverse thermal conduction coefficients; the ionization loss term on the left-hand side (LHS) of (1.27) may become significant when electrons are heated up to high temperature enabling thermal ionization.

Problems

P1.1. Show that the electric field is transformed away in a moving frame with a velocity $V_E = (\mathbf{E} \times \mathbf{B})/B^2$, where $\mathbf{E}$ and $\mathbf{B}$ are uniform static fields and perpendicular to each other.

P1.2. Debye length: if singly charged ion plasma has a finite temperature T_i, show that the Debye length is given to be

$$\lambda_D = (\lambda_{De}^{-2} + \lambda_{Di}^{-2})^{-\frac{1}{2}}$$

where the electron/ion Debye lengths $\lambda_{De,i} = (\varepsilon_0 T_{e,i}/n_0 e^2)^{\frac{1}{2}}$. [hint: the Poisson equation is modified to be $\nabla^2\varphi(r) = -(q_0/\varepsilon_0)\delta(r) + e(\delta n_e - \delta n_i)/\varepsilon_0$]

P1.3. Calculate the Debye length of ionospheric plasma, where the plasma density $n_0 = 4 \times 10^{11}$ m^{-3}, $T_e = 1200$ K, and $T_i = 1000$ K.

P1.4. Show that the electron plasma frequency $f_{pe} = 9 \times \sqrt{n_0}$ Hz, where n_0 is number of electrons per cubic meter.

P1.5. Determine the condition that Eq. (1.16a) is a solution of Eq. (1.14). What is the terminology of this condition?

P1.6. Show that $f_a(v_\perp, v_\parallel)$ is a general stationary (steady state) solution of the Boltzmann (Vlasov) equation (1.17) in the absence of external perturbations, where $v_\perp = |\mathbf{v}_\perp|$ and $v_\parallel = |\mathbf{v}_\parallel|$ are the magnitude of the velocity components perpendicular and parallel to the background uniform magnetic field.

P1.7. A statistical approach to derive the Vlasov equation (1.17). Consider an N particle system, each particle has a trajectory $\mathbf{X}_j = (\mathbf{r}_j, \mathbf{v}_j)$ which is governed by the trajectory equations: $d\mathbf{r}_j(t)/dt = \mathbf{v}_j(t)$ and $d\mathbf{v}_j(t)/dt = \mathbf{F}_{aj}/m_a = (q_a/m_a)[\mathbf{E}(\mathbf{r}_j, t) + \mathbf{v}_j \times \mathbf{B}_0]$, where the initial condition $\mathbf{X}_j(0) = (\mathbf{r}_{j0} = \mathbf{r}_j(0), \mathbf{v}_{j0} = \mathbf{v}_j(0))$ is a random variable. Assume that each particle has a point volume, its density is represented by a delta function, e.g., $\delta(\mathbf{X} - \mathbf{X}_i) = \delta(\mathbf{r} - \mathbf{r}_i)\delta(\mathbf{v} - \mathbf{v}_i)$, where $\mathbf{X} = (\mathbf{r}, \mathbf{v})$; thus, the density distribution of the system is given by the Klimontovich distribution function:

$$\mathrm{N}(\mathbf{X}; \{\mathbf{X}_\mathrm{j}\}) = \sum_{j=1}^{N} \delta(\mathbf{X} - \mathbf{X}_\mathrm{j}), \quad \text{where}$$

$$\{\mathbf{X}_\mathrm{j}\} = \{\mathbf{X}_1, \mathbf{X}_2, \ldots, \mathbf{X}_\mathrm{N}\}$$

which poses a microstate of this N particle system and represents a point in 6N-dimensional phase space (the Γ space). A distribution of points in the Γ space stands for an ensemble of similar systems in different microstates.

(1) Show that $\mathrm{N}(\mathbf{X}; \{\mathbf{X}_\mathrm{j}\})$ is governed by the Klimontovich equation: $\partial_\mathrm{t}\mathrm{N} + \mathbf{v} \cdot \boldsymbol{\nabla}_\mathbf{r}\mathrm{N} + \mathbf{a} \cdot \boldsymbol{\nabla}_\mathbf{v}\mathrm{N} = 0$, where the acceleration $\mathbf{a} = \mathrm{d}\mathbf{v}/\mathrm{dt} = (\mathrm{q_a}/\mathrm{m_a})[\mathbf{E}(\mathbf{r}, \mathrm{t}) + \mathbf{v} \times \mathbf{B}_0]$.

(2) Set $\mathrm{f_a}(\mathbf{r}, \mathbf{v}, \mathrm{t}) = \langle \mathrm{N} \rangle$, where $\langle\ \rangle$ stands for ensemble average; derive the Vlasov equation from the Klimontovich equation.

P1.8. The pressure of a gas is defined to be the force exerted by the gas (through collisions) on a unit surface area. In each elastic collision, the momentum change of the gas particle is $\Delta\mathbf{p} = -2\mathrm{m_a}\mathbf{v}_\perp$, where $\mathbf{v}_\perp$ is component of the particle velocity normal to the surface.

An ideal gas has the density distribution given by Eq. (1.19). Consider an imaginary wall on the y–z plane as shown in Fig. P1.1, particles of velocity $\mathbf{v}$ in the rectangular column will collide this wall during a 1-second period. With the aid of this figure and the calculated force on this wall, show that the pressure of the gas is obtained to be $\mathrm{P_a} = \mathrm{n_0T_a}$.

[Hint: Apply this relation $\int_0^\infty \mathrm{e}^{-\mathrm{x}^2/2}\mathrm{dx} = (\pi/2)^{\frac{1}{2}}$; this integration is illustrated as follows:

$$\left(\int_0^\infty \mathrm{e}^{-\mathrm{x}^2/2}\mathrm{dx}\right)\left(\int_0^\infty \mathrm{e}^{-\mathrm{y}^2/2}\mathrm{dy}\right) = \iint_0^\infty \mathrm{e}^{-(\mathrm{x}^2+\mathrm{y}^2)/2}\mathrm{dx\,dy}$$

We now change coordinate system from the rectangular coordinates (x, y) to the polar coordinates (r, θ). With the

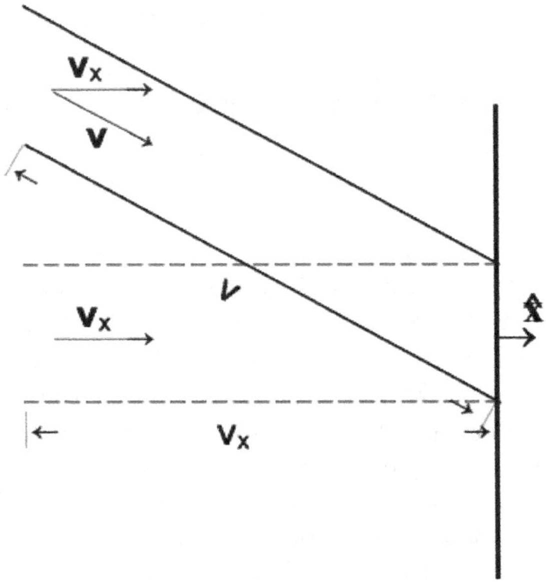

Fig. P1.1. Schematic used to calculate the total momentum change in one second of gas particles due to elastic reflection from a virtual wall of unit area.

aid of $x^2 + y^2 = r^2$ and $dx\, dy = rdr\, d\theta = \frac{1}{2}dr^2 d\theta$, it gives

$$\iint_0^{\infty} e^{-(x^2+y^2)/2} dx\, dy = \int_0^{\pi/2} d\theta \int_0^{\infty} e^{-r^2/2} d(r^2/2) = \frac{\pi}{2}\Bigg]$$

P1.9. Equation (1.19) is an isotropic velocity distribution. It can be extended to define a speed distribution $f_{as}(v)$ and an energy distribution $F_a(\varepsilon)$, where $v = |\mathbf{v}|$ and $\varepsilon = \frac{1}{2}\, m_a v^2$, through the identities

$$n_0 = \int f_{a0}(\mathbf{v}) d\mathbf{v} = \int_0^{\infty} f_{as}(v) dv = \int_0^{\infty} F_a(\varepsilon) d\varepsilon$$

In the isotropic situation, $\int f_{a0}(\mathbf{v}) d\mathbf{v} = 4\pi \int_0^{\infty} v^2 f_{a0}(v) dv$; thus, by comparison, it leads to

$$f_{as}(v) = 4\pi v^2 f_{a0}(v) = n_0 \left(\frac{2}{\pi}\right)^{\frac{1}{2}} (v^2/v_{ta}^3) \exp\left(-\frac{v^2}{2v_{ta}^2}\right) \tag{P1.1}$$

and

$$F_a(\varepsilon) = n_0 \left(\frac{2}{\sqrt{\pi}}\right) T_a^{-3/2} \varepsilon^{\frac{1}{2}} \exp\left(-\frac{\varepsilon}{T_a}\right) \tag{P1.2}$$

(1) Apply (P9.1) to show that the average speed of the Maxwellian (1.19) is $\langle v \rangle = (8T_a/\pi m_a)^{\frac{1}{2}}$.

(2) The most probable speed is the speed most likely to be possessed by any particle (of the same mass *m*) in the gas and corresponds to the maximum value of $f_{as}(v)$. Apply (P9.1) to show that the most probable speed $v_p = (2T_a/m_a)^{\frac{1}{2}}$.

(3) The thermal energy of a gas particle is the average of its kinetic energy $\varepsilon = \frac{1}{2} m_a v^2$. Apply (P9.2) to show that the thermal energy of a particle in an ideal gas is $\langle \varepsilon \rangle = 3T_a/2$. A root mean square (rms) speed v_{rms} is then determined through $\frac{1}{2} m_a v_{rms}^2 = 3T_a/2$ to be $v_{rms} = (3T_a/m_a)^{\frac{1}{2}}$.

P1.10. An electron is situated in a uniform magnetic field $\mathbf{B}_0 = \hat{\mathbf{z}} B_0$. At $t = 0$, an ac electric field $\mathbf{E} = \hat{\mathbf{x}} E_0 \cos \omega t$ is switched on. Determine the electron motion if:

(1) At $t = 0$, the electron was at rest.

(2) At $t = 0$, the electron was gyrating in the magnetic field at the velocity $\mathbf{v}(t = 0) = (\hat{\mathbf{x}} + \hat{\mathbf{y}})(v_0/\sqrt{2})$.

(3) Prove that the solution of (2) is the superposition of the gyration and the motion obtained in (1).

P1.11. Use Eq. (1.17) to show that in the absence of collisions the temperature T_a is governed by the equation

$$\frac{\partial T_a}{\partial t} + \mathbf{v}_a \cdot \boldsymbol{\nabla} T_a + \left(\frac{2T_a}{3}\right) \boldsymbol{\nabla} \cdot \mathbf{v}_a = 0$$

[Hint: use (1.17) to derive the moment equation of $|\mathbf{v}|^2$]

P1.12. Show that Eq. (1.22b) can be written as $\boldsymbol{\nabla} \times \mathbf{H} = \mathbf{J}_e + \partial \mathbf{D}/\partial t$ in the form of Ampere's Law, where the magnetic field intensity H is related to the magnetic flux density B by $H = B/\mu_0$.

P1.13. Design a Plasma Rocket Engine shown in Fig. P1.2(a).

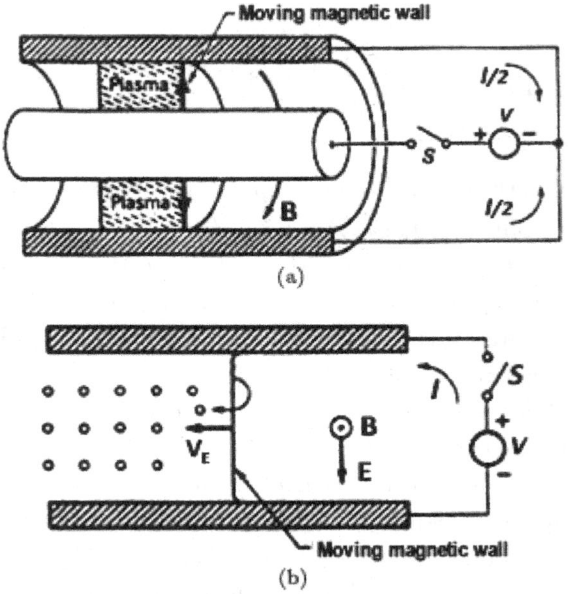

Fig. P1.2. (a) Cut view of a plasma rocket engine and (b) a two-dimensional plane model.

It is desired to design a plasma rocket engine using the plane model shown in Fig. P1.2(b) to develop a thrust: $T = m_A v_f = 0.2$ N and a power: $P = m_A v_f^2/2 = 0.1$ MW, where m_A is the mass of the fuel (cesium, $^{55}C_S^{133}$) used in 1 second and v_f is its exhaust velocity with respect to the rocket.

Determine

(1) The applied voltage V

(2) N, number of C_S^+ ions ejected per second (C_S^+ is singly ionized)

Consider an ideal case, namely, there is no internal power loss so that the output power of the voltage source equals the power of the engine. You have to choose the dimension of the engine first and choose the operating mode, i.e., how many fuel injections per second will be run.

Chapter 2

Electromagnetic Property of Plasma and Plasma Modes

Plasma is a dielectric medium, which supports three types of waves including electromagnetic (EM), electrostatic (ES), as well as hybrid:

(1) EM wave is a transverse wave (i.e., TEM); it carries two fields, electric field $\mathbf{E}$ and magnetic field $\mathbf{H}$, which are perpendicular to each other and also to the wave propagation direction referred by the wavevector $\mathbf{k}$, i.e., $\mathbf{E} \perp \mathbf{H} \perp \mathbf{k}$, where $\mathbf{k} = \hat{\mathbf{k}}(2\pi/\lambda)$ and λ is the wavelength; the wavenumber $\mathrm{k} = 2\pi/\lambda$ infers the number of waves in unit length; thus $\boldsymbol{\nabla} \cdot \mathbf{E} = 0$, i.e., $\mathrm{n_e} = \mathrm{n_i} = \mathrm{n_0}$, there is no space charge associated with EM waves. However, the wave electric field will drive a current in plasma. Thus, TEM wave is governed only by the two curl equations in Maxwell's equations. The induced current density $\mathbf{J}$ appearing in the curl equation of the Ampere's law modifies the wave fields self-consistently as well as the wave propagation properties.

(2) ES wave is a longitudinal wave; it carries only electric field $\mathbf{E}$, which is parallel to the wave propagation direction, i.e., $\mathbf{E}||\mathbf{k}$, and thus $\mathbf{H} = 0$. Therefore, the wave is governed only by the divergence equation (Gauss's law) for the electric field. In this case, a scalar potential ϕ can be introduced to define $\mathbf{E} = -\boldsymbol{\nabla}\phi$, which converses the divergence equation to the Possion's equation $\nabla^2\phi = -\mathrm{e}(\mathrm{n_i} - \mathrm{n_e})/\varepsilon_0$.

(3) Hybrid wave is neither transverse nor longitudinal and exists in magneto plasma; the wave electric field has an angle other than

$0°$ or $90°$ with respect to the propagation direction. Generally, the full set of Maxwell's equations is needed to specify the wave; some simplification, based on its nature tending to be EM or ES, may be introduced. In the high-frequency domain, ions are immobile, only electron plasma responds to the wave fields; on the other hand, quasi-neutral condition in the density perturbations may be assumed in the low-frequency domain. Hybrid waves are also plasma modes.

A mode of a linear system is an oscillation that can self-sustain in the system. For example, apply an impulse to a lossless system; oscillations at all frequencies will be excited. However, not all of the oscillations can persist in time; in fact, most of the oscillations will damp away via phase mixing (among themselves or caused by the boundary effects). In the steady state, only a few oscillations may remain, and these lasting oscillations are the "modes" of the system. These are "eigen-modes". In practice, there are always some losses in the system; in this case, these lasting oscillations are "quasi-modes" which are also damped in time (but much slower). In the mathematical analysis, modes represent the non-trivial space–time harmonic solutions of the source-free governing equations of a linear system. The condition that non-trivial solutions of the source-free governing equations exist sets up a dispersion equation. Each solution $\omega(\mathrm{k})$ or $\mathrm{k}(\omega)$ of the dispersion equation is the dispersion relation of a branch of mode, which can be plotted on a $\mathrm{k} - \omega$ or $\omega - \mathrm{k}$ plane as the dispersion curve of this branch of mode. Each point (k, ω) or (ω, k) on the curve defines the wavenumber k and frequency ω of a mode in this branch.

2.1 Plasma Dielectric Tensor and Dispersion Equation

Wave propagation in plasma is subjected to the conditions imposed by the EM property of the plasma, which is characterized by a dielectric tensor. Source-free Maxwell's equations determine this dielectric tensor in the $\mathrm{k} - \omega$ space for the space–time harmonic fields and the responding physical quantities, which all have the function form $\exp[i(k\cdot r-\omega t)]$, e.g., $\mathbf{A} = \tilde{\mathbf{A}}\exp[i(k\cdot r-\omega t)]$, where $\tilde{\mathbf{A}}$ is constant amplitude. Thus, the differential operators $\partial/\partial t$ and $\boldsymbol{\nabla}$ can be replaced by $-i\omega$ and $i\mathbf{k}$ to convert differential equations to algebraic equations, in which the common factor $\exp[i(k\cdot r-\omega t)]$ in every term of the equations can be factored out and removed. In other words, all the differential equations for the fields and physical quantities "$\mathbf{A}$" are reduced to algebraic equations for the corresponding amplitudes "$\tilde{\mathbf{A}}$". In the subsequent discussion, "$\tilde{\mathbf{A}}$" in the algebraic equations will be replaced by "$\mathbf{A}$", the same symbols of the fields and physical quantities, to simplify the expression.

In the source-free situation, (1.22) and (1.23) become

$$\mathbf{k}\times\mathbf{E} = \omega\mathbf{B} \tag{2.1a}$$

$$\mathbf{k}\times\mathbf{B} + \frac{\omega}{c^2}\mathbf{E} = -i\mu_0\mathbf{J} \tag{2.1b}$$

$$\mathbf{k}\cdot\mathbf{E} = -i\frac{\rho}{\varepsilon_0} \tag{2.1c}$$

$$\mathbf{k}\cdot\mathbf{B} = 0 \tag{2.1d}$$

and

$$\rho = \frac{\mathbf{k}\cdot\mathbf{J}}{\omega} = \frac{\mathbf{k}\cdot\overleftrightarrow{\boldsymbol{\sigma}}\cdot\mathbf{E}}{\omega} \tag{2.2}$$

where a conductivity tensor $\overleftrightarrow{\boldsymbol{\sigma}}(\mathbf{k},\omega)$ is introduced to set the relationship $\mathbf{J} = \overleftrightarrow{\boldsymbol{\sigma}}\cdot\mathbf{E}$. In the $k-\omega$ space, the electric flux density $\mathbf{D}$ and the electric field $\mathbf{E}$ are related through a dielectric tensor $\overleftrightarrow{\boldsymbol{\varepsilon}}(\mathbf{k},\omega)$ to be

$$\mathbf{D} = \varepsilon_0\mathbf{E} + \mathcal{P} = \varepsilon_0\overleftrightarrow{\boldsymbol{\varepsilon}}\cdot\mathbf{E} \tag{2.3}$$

where $\mathbf{k}\cdot\mathcal{P} = i\rho$.

2.1.1 *Relationship of the plasma dielectric tensor and the conductivity tensor*

With the aid of (2.2) and (2.1c), $\mathbf{k}\cdot\mathbf{D}$ of (2.3) becomes

$$\varepsilon_0\mathbf{k}\cdot\left[\overleftrightarrow{\mathbf{I}} + i\frac{\overleftrightarrow{\boldsymbol{\sigma}}}{\omega\varepsilon_0}\right]\cdot\mathbf{E} = 0 = \varepsilon_0\mathbf{k}\cdot\overleftrightarrow{\boldsymbol{\varepsilon}}\cdot\mathbf{E} \tag{2.4}$$

which leads to the relationship between the plasma dielectric tensor $\overleftrightarrow{\varepsilon}(\mathbf{k},\omega)$ and the conductivity tensor $\overleftrightarrow{\sigma}(\mathbf{k},\omega)$ to be

$$\overleftrightarrow{\varepsilon}(\mathbf{k},\omega) = \overleftrightarrow{\mathbf{I}} + \mathrm{i}\frac{\overleftrightarrow{\sigma}(\mathbf{k},\omega)}{\omega\varepsilon_0} \tag{2.5}$$

where $\overleftrightarrow{\mathbf{I}} = \hat{\mathbf{x}}\hat{\mathbf{x}} + \hat{\mathbf{y}}\hat{\mathbf{y}} + \hat{\mathbf{z}}\hat{\mathbf{z}}$ is an unit operator dyadic. As shown, the plasma dielectric tensor can be determined through the conductivity tensor, which is the transfer function of the current density induced by a plane wave (space–time harmonic) field in plasma.

2.1.1.1 *Plasma dispersion equations for electromagnetic/hybrid modes and for electrostatic modes*

With the aid of (2.5), (2.1a) and (2.1b) are combined to be

$$\left[\overleftrightarrow{\varepsilon}(\mathbf{k},\omega) - \left(\frac{\mathrm{kc}}{\omega}\right)^2 \overleftrightarrow{\mathbf{I}}_{\mathrm{T}}\right] \cdot \mathbf{E} = 0 \tag{2.6}$$

where $\overleftrightarrow{\mathbf{I}}_{\mathrm{T}} = \overleftrightarrow{\mathbf{I}} - \hat{\mathbf{k}}\hat{\mathbf{k}} = \overleftrightarrow{\mathbf{I}} - \overleftrightarrow{\mathbf{I}}_{\mathrm{L}}$ is a transverse unit operator dyadic, where $\hat{\mathbf{k}} = \mathbf{k}/\mathrm{k}$ is a unit vector in the direction of $\mathbf{k}$ and $\overleftrightarrow{\mathbf{I}}_{\mathrm{L}} = \hat{\mathbf{k}}\hat{\mathbf{k}}$ represents a longitudinal unit operator dyadic.

Equation (2.6) possesses non-trivial solutions if the determinant of the tensor operator acting on $\mathbf{E}$ is zero; it is expressed to be

$$\det\left|\left[\overleftrightarrow{\varepsilon}(\mathbf{k},\omega) - \left(\frac{\mathrm{kc}}{\omega}\right)^2 \overleftrightarrow{\mathbf{I}}_{\mathrm{T}}\right]\right| = 0 \tag{2.7}$$

This equation sets up a dispersion equation, which determines the frequency-wavevector dispersion relations for all branches of modes (self-sustain oscillations) in plasma. However, (2.7) is mainly applied for describing the EM modes. In general, the electron thermal speed is much lower than the speed of the EM modes; thus the thermal terms in $\overleftrightarrow{\varepsilon}(\mathbf{k},\omega)$ may be neglected in analyzing (2.7).

In the ES case, $\mathbf{E} = -\mathrm{i}\mathbf{k}\phi$, the dispersion equation (2.7) is reduced to

$$\hat{\mathbf{k}} \cdot \overleftrightarrow{\varepsilon}(\mathbf{k},\omega) \cdot \hat{\mathbf{k}} = \varepsilon_{\mathrm{L}}(\mathbf{k},\omega) = 1 + \mathrm{i}\left(\frac{\hat{\mathbf{k}} \cdot \overleftrightarrow{\sigma}(\mathbf{k},\omega) \cdot \hat{\mathbf{k}}}{\omega\varepsilon_0}\right) = 0 \tag{2.8}$$

but the lowest order thermal terms in $\overleftrightarrow{\boldsymbol{\sigma}}(\mathbf{k},\omega)$ have to be retained. It will be shown in Chapter 3 that $\varepsilon_\mathrm{L}(\mathbf{k},\omega)$, derived from the kinetic approach, is a complex function even in a collisionless plasma. Let, $(\mathbf{k}_0, \omega_0 + \mathrm{i}\gamma)$ be a solution of the dispersion equation $\varepsilon_\mathrm{L}(\mathbf{k},\omega) = 0$, where $\mathbf{k}_0$ and ω_0 are the wavevector and real frequency of a plasma mode; thus $\varepsilon_\mathrm{L}(\mathbf{k}_0, \omega_0 + \mathrm{i}\gamma) = 0 = \varepsilon_\mathrm{L}(\mathbf{k}_0,\omega_0) + \mathrm{i}\gamma\partial\varepsilon_\mathrm{L}(\mathbf{k}_0,\omega)/\partial\omega|\omega_0 + \cdots \cong \varepsilon_\mathrm{Lr}(\mathbf{k}_0,\omega_0) + \mathrm{i}\,\varepsilon_\mathrm{Li}(\mathbf{k}_0,\omega_0) + \mathrm{i}\gamma\partial\varepsilon_\mathrm{Lr}(\mathbf{k}_0,\omega_0)/\partial\omega_0$, where $\varepsilon_\mathrm{Lr}(\mathbf{k}_0,\omega_0)$ and $\varepsilon_\mathrm{Li}(\mathbf{k}_0,\omega_0)$ are the real and imaginary part of $\varepsilon_\mathrm{L}(\mathbf{k}_0,\omega_0)$, and $|\gamma| \ll \omega_0$ is assumed. Therefore, $(\mathbf{k}_0,\omega_0)$ is governed by the dispersion equation

$$\varepsilon_\mathrm{Lr}(\mathbf{k}_0,\omega_0) = 0 \tag{2.9}$$

and the damping rate of the mode is given to be

$$\gamma = -\frac{\varepsilon_\mathrm{Li}(\mathbf{k}_0,\omega_0)}{\left[\frac{\partial\varepsilon_\mathrm{Lr}(\mathbf{k}_0,\omega_0)}{\partial\omega_0}\right]} \tag{2.10}$$

2.2 Cutoff and Resonance

EM wave propagates at a phase velocity $\mathrm{v_p} = \omega/\mathrm{k}$, where ω and k are related through a dispersion relation determined by (2.7). If there is a frequency band that k becomes imaginary, the EM wave is cutoff from propagation in this frequency band. In other words, in the propagation band, $(\mathrm{kc}/\omega)^2 > 0$ and in the cutoff band, $(\mathrm{kc}/\omega)^2 < 0$. There are two situations in the sign change of $(\mathrm{kc}/\omega)^2$. One is going through $(\mathrm{kc}/\omega)^2 = 0$, i.e., $\mathrm{k} = 0$. The frequency at $\mathrm{k} = 0$ (i.e., $\mathrm{v_p} \to \infty$) is defined to be the cutoff frequency ω_c of this branch of EM modes. The second one is going through $(\mathrm{kc}/\omega)^2 \to \infty$, i.e., $\mathrm{k} \to \infty$. As $\mathrm{v_p} \to 0$, wave interacts strongly with the plasma, it is called resonance. The frequency at $\mathrm{k} \to \infty$ (i.e., $\mathrm{v_p} \to 0$) is called a resonance frequency ω_R.

2.3 Derivation of the Dielectric Tensor — Fluid Approach

Set, $\mathrm{n_a} = \mathrm{n_0} + \delta\mathrm{n_a}$, $\mathbf{v}_\mathrm{a} = \delta\mathbf{v}_\mathrm{a}$ and $\mathbf{B} = \hat{\mathbf{z}}\,\mathrm{B_0}$, where the subscript "a" = e, i; "δ" signifies "first order (linear)" responses of plasma to the wave field $\mathbf{E}$, the continuity equation (1.20) and the momentum equation (1.21) are linearized (with respective to "δ") in the $k - \omega$ space to be

$$\frac{\delta n_a}{n_0} = \frac{\mathbf{k} \cdot \delta \mathbf{v}_a}{\omega} \tag{2.11}$$

and

$$(\omega \delta \mathbf{v}_a - i\Omega_a \delta \mathbf{v}_a \times \hat{\mathbf{z}}) = \mathbf{k} \left(\frac{\gamma_a T_a}{m_a}\right) \left(\frac{\delta n_a}{n_0}\right) + \frac{i q_a \mathbf{E}}{m_a} \tag{2.12}$$

Substituting (2.11) into (2.12), yields

$$\left[\overleftrightarrow{\mathbf{I}} + i\left(\frac{\Omega_a}{\omega}\right) \hat{\mathbf{z}} \times \overleftrightarrow{\mathbf{I}} - \left(\frac{\gamma_a v_{ta}^2}{\omega^2}\right) \mathbf{kk}\right] \cdot \delta \mathbf{v}_a = i \frac{q_a \mathbf{E}}{m_a \omega} \tag{2.13}$$

In carrying out the inversion of the matrix $[\overleftrightarrow{\mathbf{I}} + i(\Omega_a/\omega)\hat{\mathbf{z}} \times \overleftrightarrow{\mathbf{I}} - (\gamma_a v_{ta}^2/\omega^2)\mathbf{kk}]$ on the left-hand side (LHS) of (2.13), we set $\mathbf{k} = \hat{\mathbf{x}}\, k_\perp + \hat{\mathbf{z}}\, k_{||}$ without losing the generality. We then obtain

$$\delta \mathbf{v}_a = i\left(\frac{q_a}{m_a \omega}\right) \overleftrightarrow{\boldsymbol{\mu}}_a \cdot \mathbf{E} \tag{2.14}$$

where the inverted matrix (dyadic) is given to be

$$\begin{aligned} \overleftrightarrow{\boldsymbol{\mu}}_a(\mathbf{k}, \omega) = \Bigg\{ & \hat{\mathbf{x}}\hat{\mathbf{x}} + i\left(\frac{\Omega_a}{\omega}\right)(\hat{\mathbf{x}}\hat{\mathbf{y}} - \hat{\mathbf{y}}\hat{\mathbf{x}}) + \left(1 - \alpha_a \frac{k_\perp^2}{k^2}\right)\hat{\mathbf{y}}\hat{\mathbf{y}} \\ & + \alpha_a \frac{k_\perp k_{||}}{k^2}(\hat{\mathbf{x}}\hat{\mathbf{z}} + \hat{\mathbf{z}}\hat{\mathbf{x}}) - i\left(\frac{\Omega_a}{\omega}\right)\alpha_a \frac{k_\perp k_{||}}{k^2}(\hat{\mathbf{y}}\hat{\mathbf{z}} - \hat{\mathbf{z}}\hat{\mathbf{y}}) \\ & + \left[\left(1 - \frac{\Omega_a^2}{\omega^2}\right)\left(1 + \alpha_a \frac{k_{||}^2}{k^2} - \alpha_a \frac{k_\perp^2}{k^2}\right)\right]\hat{\mathbf{z}}\hat{\mathbf{z}} \Bigg\} \\ & \Big/ \left[\left(1 - \frac{\Omega_a^2}{\omega^2}\right) - \alpha_a \frac{k_\perp^2}{k^2}\right] \end{aligned} \tag{2.15a}$$

and $\alpha_a = (\gamma_a k^2 v_{ta}^2/\omega^2)/(1 - \gamma_a k_{||}^2 v_{ta}^2/\omega^2)$.

In cold plasma, $\alpha_a = 0$ and (2.15a) reduces to

$$\overleftrightarrow{\boldsymbol{\mu}}_a(\mathbf{k},\omega) = \left[\hat{\mathbf{x}}\hat{\mathbf{x}} + i\left(\frac{\Omega_a}{\omega}\right)(\hat{\mathbf{x}}\hat{\mathbf{y}} - \hat{\mathbf{y}}\hat{\mathbf{x}}) + \hat{\mathbf{y}}\hat{\mathbf{y}} + \left(1 - \frac{\Omega_a^2}{\omega^2}\right)\hat{\mathbf{z}}\hat{\mathbf{z}}\right] \Big/ \left(1 - \frac{\Omega_a^2}{\omega^2}\right) \tag{2.15b}$$

With the aid of (2.14), the current density driven by wave electric field is derived to be

$$\mathbf{J} = \sum_{a=e,i} q_a n_0 \delta\mathbf{v}_a = i\varepsilon_0 \sum_{a=e,i} \left(\frac{\omega_{pa}^2}{\omega}\right) \overleftrightarrow{\boldsymbol{\mu}}_a \cdot \mathbf{E} \tag{2.16}$$

where $\omega_{pa} = (n_0 q_a^2/m_a\varepsilon_0)^{1/2}$ is the plasma frequency of species "a". Thus the conductivity tensor is determined to be

$$\overleftrightarrow{\boldsymbol{\sigma}}(\mathbf{k},\omega) = i\varepsilon_0 \sum_{a=e,i} \left(\frac{\omega_{pa}^2}{\omega}\right) \overleftrightarrow{\boldsymbol{\mu}}_a(\mathbf{k},\omega) \tag{2.17}$$

2.3.1 *Plasma dielectric tensor*

Substitute (2.17) into (2.5), the plasma dielectric tensor is obtained as

$$\overleftrightarrow{\boldsymbol{\varepsilon}}(\mathbf{k},\omega) = \overleftrightarrow{\mathbf{I}} - \sum_{a=e,i} \left(\frac{\omega_{pa}}{\omega}\right)^2 \overleftrightarrow{\boldsymbol{\mu}}_a(\mathbf{k},\omega) \tag{2.18a}$$

where $\overleftrightarrow{\boldsymbol{\mu}}_a(\mathbf{k},\omega)$ is given by (2.15a).

In cold plasma, with the aid of (2.15b), (2.18a) becomes

$$\overleftrightarrow{\boldsymbol{\varepsilon}}(\mathbf{k},\omega) = \varepsilon_1\hat{\mathbf{x}}\hat{\mathbf{x}} + i\varepsilon_2(\hat{\mathbf{x}}\hat{\mathbf{y}} - \hat{\mathbf{y}}\hat{\mathbf{x}}) + \varepsilon_1\hat{\mathbf{y}}\hat{\mathbf{y}} + \varepsilon_3\hat{\mathbf{z}}\hat{\mathbf{z}} \tag{2.18b}$$

where

$$\varepsilon_1 = 1 - \sum_{a=e,i} \frac{\omega_{pa}^2}{\omega^2 - \Omega_a^2}, \quad \varepsilon_2 = -\sum_{a=e,i} \frac{\Omega_a\omega_{pa}^2}{\omega(\omega^2 - \Omega_a^2)},$$

$$\varepsilon_3 = 1 - \sum_{a=e,i} \left(\frac{\omega_{pa}}{\omega}\right)^2.$$

2.3.2 *Dispersion equation of electromagnetic/hybrid modes in cold plasma*

Set $\hat{\mathbf{k}} = (k_\perp/k)\hat{\mathbf{x}} + (k_{||}/k)\hat{\mathbf{z}}$, then $\overleftrightarrow{\mathbf{I}}_T = (k_{||}/k)^2\hat{\mathbf{x}}\hat{\mathbf{x}} + \hat{\mathbf{y}}\hat{\mathbf{y}} - (k_{||}k_\perp/k^2)(\hat{\mathbf{x}}\,\hat{\mathbf{z}} + \hat{\mathbf{z}}\,\hat{\mathbf{x}}) + (k_\perp/k)^2\hat{\mathbf{z}}\,\hat{\mathbf{z}}$, and the dispersion equation (2.7) for a cold plasma is expressed explicitly to be

$$\left[\varepsilon_3 - \left(\frac{k_\perp c}{\omega}\right)^2\right]\left[\varepsilon_1 + \varepsilon_2 - \left(\frac{k_{||}c}{\omega}\right)^2\right]\left[\varepsilon_1 - \varepsilon_2 - \left(\frac{k_{||}c}{\omega}\right)^2\right]$$
$$- \left(\frac{k_\perp c}{\omega}\right)^2 \left\{\varepsilon_1\left[\varepsilon_3 - \left(\frac{k_\perp c}{\omega}\right)^2\right]\right.$$
$$\left. - \left(\frac{k_{||}c}{\omega}\right)^2\left[\left(\frac{k_{||}c}{\omega}\right)^2 + \sum_{a=e,i}\frac{\omega_{pa}^2\Omega_a^2}{\omega^2(\omega^2 - \Omega_a^2)}\right]\right\} = 0 \qquad (2.19)$$

Equation (2.19) can describe properly the propagation characteristics of the EM modes and hybrid modes dominated with EM nature. However, the thermal terms in (2.15a) have to be retained for hybrid modes dominated with ES nature and the ES modes.

2.3.3 *Dispersion equation of electrostatic modes*

The dispersion properties of the ES modes are characterized directly by the dispersion equation (2.8) which is expressed explicitly to be

$$1 - \sum_{a=e,i}\left\{\frac{\left(\frac{\omega_{pa}}{\omega}\right)^2\left[1 - \left(\frac{k_{||}}{k}\right)^2\left(\frac{\Omega_a}{\omega}\right)^2\right]}{\left(1 - \frac{\Omega_a^2}{\omega^2} - \frac{\alpha_a k_\perp^2}{k^2}\right)\left(1 - \frac{\gamma_a k_{||}^2 v_{ta}^2}{\omega^2}\right)}\right\} = 0 \qquad (2.20)$$

where $v_{ta} = (T_a/m_a)^{1/2}$.

2.4 Branches of Electrostatic Plasma Modes

2.4.1 *Plasma modes in unmagnetized plasma*

Set $\Omega_a = 0$, (2.20) is reduced to

$$1 - \sum_{a=e,i} \frac{\omega_{pa}^2}{\omega^2 - \gamma_a k^2 v_{ta}^2} = 0 \tag{2.21}$$

In the following, (2.21) is analyzed for two branches of ES modes in high- and low-frequency domains, respectively.

2.4.1.1 *High-frequency domain, $\omega \geq \omega_{pe}$*

Neglecting ion terms in (2.21), leads to $1 - [\omega_{pe}^2/(\omega^2 - 3k^2 v_{te}^2)] = 0$, where $\gamma_e = 3$ for one-dimensional oscillation. The dispersion relation of electron plasma (Langmuir) modes is obtained to be

$$\omega(k) = (\omega_{pe}^2 + 3k^2 v_{te}^2)^{1/2} = \omega_{ek} \tag{2.22}$$

This is the same result as that presented in Sec. 1.1.2 of Chapter 1 to demonstrate the collective phenomenon. It shows that the electron plasma wave oscillates with frequency $\omega \geq \omega_{pe}$ and propagates at phase velocity $v_p = \sqrt{3}[\omega/(\omega^2 - \omega_{pe}^2)^{1/2}]v_{te}$.

2.4.1.2 *Low-frequency domain, $\omega^2 \ll k^2 v_{te}^2$*

Equation (2.21) becomes $1 + k_{De}^2/k^2 - \omega_{pi}^2/(\omega^2 - 3k^2 v_{ti}^2) = 0$, where $k_{De} = 1/\lambda_{De} = \omega_{pe}/v_{te}$ and λ_{De} is the Debye length introduced in Sec. 1.1 of Chapter 1; $\gamma_e = 1$ for isothermal compression and $\gamma_i = 3$ for one-dimensional adiabatic compression. For $k^2 \ll k_{De}^2$, the dispersion relation of the ion acoustic wave is obtained to be

$$\omega(k) = kC_s = \omega_{sk} \tag{2.23}$$

where the ion acoustic speed $C_s = [(T_e + 3T_i)/m_i]^{1/2}$. Normally, $T_e \gg T_i$, the ion acoustic speed is proportional to the electron temperature and inversely proportional to the ion mass. This is another example of the collective effect; because electron moves much faster, the motion of the ion density perturbation is mainly through the action of electron shielding on the induced electric field of the ion density perturbation.

2.4.2 *Plasma modes in magneto plasma*

2.4.2.1 *High-frequency domain,* $\omega \geq \omega_{pe}$

(1) Propagation parallel to the magnetic field, $k_\perp = 0$ and $k_{||} = k$

This is the same situation as that of unmagnetized plasma; thus, the dispersion relation is the same as (2.22) for electron plasma mode.

(2) Propagation transverse to the magnetic field, $k_\perp = k$ and $k_{||} = 0$

Equation (2.20) becomes $1 - \omega_{pe}^2/(\omega^2 - \Omega_e^2 - 3k^2v_{te}^2) = 0$, where $\gamma_e = 3$ for one-dimensional adiabatic compression is assumed; however, the exact value of γ_e will be determined via the kinetic approach presented in Chapter 3; the dispersion relation of the upper hybrid mode is obtained to be

$$\omega(k) = (\omega_{uH}^2 + 3k^2v_{te}^2)^{1/2} = \omega_{uk} \tag{2.24}$$

where $\omega_{uH} = (\omega_{pe}^2 + \Omega_e^2)^{1/2}$ is the upper hybrid resonance frequency, which is larger than the electron plasma frequency ω_{pe} due to the magnetic field. In other words, in the transverse direction, electron plasma resonances to the perturbation at upper hybrid resonance frequency ω_{uH}, rather than at electron plasma frequency ω_{pe}. As demonstrated in Sec. 1.1.1.2 of Chapter 1, collective electron plasma oscillation is similar to the oscillation of a harmonic oscillator. The resonance frequency is proportional to the restoring force acting on the object of the oscillator. In the case of oscillation parallel to the magnetic field, the restoring force comes from the induced electric field which is proportional to the electron density. When the oscillation is perpendicular to the magnetic field, an additional Lorentz force, which is perpendicular to the electric force, is added to the electric force in restoring the electron displacement from the equilibrium ion distribution. Consequently, the resonance frequency increases from ω_{pe} to ω_{uH}. (see Problem P2.6).

(3) Propagation oblique to the magnetic field, $k_\perp = k\sin\vartheta$ and $k_{||} = k\cos\vartheta$

Equation (2.20) becomes $1 - (\omega_{pe}/\omega)^2[1 - (\Omega_e/\omega)^2\cos^2\vartheta]/[(1 - \Omega_e^2/\omega^2)(1 - 3k^2v_{te}^2\cos^2\vartheta/\omega^2) - 3k^2v_{te}^2\sin^2\vartheta/\omega^2] = 0$, again $\gamma_e = 3$ is assumed. This dispersion equation is re-expressed as $\omega^4 - \omega_{uk}^2\omega^2 + \Omega_e^2\omega_{ek}^2\cos^2\vartheta = 0$; and the dispersion relation of oblique propagation electron plasma mode is derived to be

$$\omega(k, \vartheta) = \frac{1}{\sqrt{2}}[\omega_{uk}^2 + (\omega_{uk}^4 - 4\Omega_e^2\omega_{ek}^2\cos^2\vartheta)^{1/2}]^{1/2}$$

$$\sim (\omega_{ek}^2 + \Omega_e^2\sin^2\vartheta)^{1/2} \qquad \text{for } \Omega_e^2\sin^2\vartheta \ll \omega_{ek}^2 \tag{2.25}$$

2.4.2.2 *Intermediate-to low-frequency domain,* $\omega^2 \ll \Omega_e^2$, $k^2v_{te}^2 \ll \omega_{pe}^2$ (*i.e.*, $k^2/k_{de}^2 \ll 1$)

The dispersion equation (2.20) becomes $1 + \omega_{pe}^2(\omega^2 - \Omega_e^2\cos^2\vartheta)/[\omega^2(\Omega_e^2 + k^2v_{te}^2) - \Omega_e^2k^2v_{te}^2\cos^2\vartheta] - \omega_{pi}^2/(\omega^2 - 3k^2v_{ti}^2) = 0$, where $\gamma_e = 1$ and $\gamma_i = 3$ are adopted. This equation has real solution under the condition (1) $\cos^2\vartheta < \omega^2/\Omega_e^2 \ll 1$ and (2) $\cos^2\vartheta > (\omega^2/\Omega_e^2)[(\Omega_e^2 + k^2v_{te}^2)/k^2v_{te}^2]$. This dispersion equation is rearranged as

$$\omega^2[\omega^2 - (\omega_{LH}^2 + k^2C_s^2)] - \zeta\omega_{LH}^2(\omega^2 - k^2C_s^2) = 0 \tag{2.26}$$

where $\zeta = (m_i/m_e)\cos^2\vartheta$ and the lower hybrid resonance frequency $\omega_{LH} = [\omega_{pi}^2\Omega_e^2/(\omega_{pe}^2 + \Omega_e^2)]^{1/2} \sim (|\Omega_e|\Omega_i)^{1/2}$ for $\omega_{pe}^2 \gg \Omega_e^2$ and $\sim \omega_{pi}$ for $\Omega_e^2 \gg \omega_{pe}^2$.

(1) Lower hybrid wave with $\cos^2\vartheta = k_{||}^2/k^2 < \omega^2/\Omega_e^2 \ll 1$

This wave propagates closely perpendicular to the magnetic field. Assume $k^2C_s^2 \ll \omega_{LH}^2(1 + \zeta)$, (2.26) is solved to obtain

$$\omega = \left[\omega_{LH}^2(1 + \zeta) + \frac{k^2C_s^2}{(1 + \zeta)}\right]^{1/2} = \omega_{Lk} \tag{2.27}$$

Equation (2.27) is the dispersion relation of lower hybrid mode. Its frequency increases with the increase of $\cos^2\vartheta$. Because $m_i/m_e \gg 1$, ζ can be much larger than 1 even with small $\cos^2\vartheta$.

(2) Ion acoustic wave with $\cos^2\vartheta > (\omega^2/\Omega_e^2)[(\Omega_e^2 + k^2 v_{te}^2)/k^2 v_{te}^2]$

Equation (2.26) is solved to obtain

$$\omega = \frac{kC_s \cos\vartheta}{(\cos^2\vartheta + m_e/m_i)^{1/2}} = \omega_{sk} \tag{2.28}$$

Equation (2.28) reduces to (2.23) in parallel propagation (i.e., $\vartheta = 0$). Because $m_e/m_i \ll 1$, wave frequency decreases slowly with the increase of the oblique angle until $\cos^2\vartheta < m_e/m_i$.

2.5 Branches of Electromagnetic Modes and Hybrid Modes in Plasma

2.5.1 *Electromagnetic wave propagation in unmagnetized plasma, $\Omega_a = 0$*

In this case, $\varepsilon_1 = \varepsilon_3 = 1 - \sum_a (\omega_{pa}/\omega)^2$ and $\varepsilon_2 = 0$. We can set $k_{||} = k$ and $k_{\perp} = 0$ without losing the generality, then it reduces (2.19) to be

$$\varepsilon_1[\varepsilon_1 - (kc/\omega)^2]^2 = 0 \tag{2.29}$$

Equation (2.19) leads to the dispersion equation

$$\varepsilon_1 = (kc/\omega)^2 \tag{2.30}$$

Thus, the propagation of EM wave in unmagnetized plasma is governed by the dispersion relation

$$\omega(k) = (\omega_{pe}^2 + k^2c^2)^{1/2} \tag{2.31}$$

which has a cutoff frequency $\omega_{co} = \omega_{pe}$. Unmagnetized plasma has a relative dielectric function $\varepsilon_r = 1 - \omega_{pe}^2/\omega^2 \leq 1$.

2.5.1.1 *Relative dielectric function of unmagnetized plasma*

Consider the propagation of a linearly polarized EM wave of frequency ω in a uniform unmagnetized plasma; set the wave to propagate in the z direction (out of the paper) and the wave electric field in the x-direction (horizontal to the right-hand side). The time harmonic electric field $\mathbf{E} = \hat{\mathbf{x}}\,\mathrm{E}$ oscillates electrons, which are displaced from the equilibrium location of immobile ions by a small distance, ξ. The charge separation and oscillation induce a polarization $\mathcal{P} = -\hat{\mathbf{x}}\,\mathrm{n_0 e}\xi$ and a polarization current with $\mathbf{J} = \partial\mathcal{P}/\partial\mathrm{t}$ in the plasma. Taking a slab of plasma, it is shown to be similar to that presented in Fig. 1.1.

With the aid of (1.7), the electron displacement ξ is given by $\xi = \mathrm{eE/m_e}\omega^2$. Thus, the polarization current density induced by the wave field is obtained to be $\mathbf{J} = -\hat{\mathbf{x}}\,\mathrm{n_0 e}\partial\xi/\partial\mathrm{t} = -\varepsilon_0(\omega_{\mathrm{pe}}^2/\omega^2)\partial\mathbf{E}/\partial\mathrm{t}$. Substitute it into (1.22b), which becomes

$$\nabla \times \mathbf{H} = \mathbf{J}_{\mathrm{e}} + \varepsilon_0\left(1 - \frac{\omega_{\mathrm{pe}}^2}{\omega^2}\right)\frac{\partial\mathbf{E}}{\partial\mathrm{t}}$$

It shows that the plasma has a relative dielectric function $\varepsilon_{\mathrm{r}}(\omega) = 1 - \omega_{\mathrm{pe}}^2/\omega^2$.

Next, we study wave propagation in magneto plasma. Two special cases of wave propagating parallel and transverse to the magnetic field are considered.

2.5.2 *Propagation parallel to the magnetic field, $\mathbf{k}_\perp = 0$ and $\mathbf{k}_{||} = \mathbf{k}$*

Equation (2.19) becomes

$$\left(1 - \sum_{\mathrm{a=e,i}}\frac{\omega_{\mathrm{pa}}^2}{\omega^2}\right)\left[1 - \sum_{\mathrm{a=e,i}}\frac{\omega_{\mathrm{pa}}^2}{\omega(\omega - \Omega_{\mathrm{a}})} - \left(\frac{\mathrm{kc}}{\omega}\right)^2\right]$$
$$\times\left[1 - \sum_{\mathrm{a=e,i}}\frac{\omega_{\mathrm{pa}}^2}{\omega(\omega + \Omega_{\mathrm{a}})} - \left(\frac{\mathrm{kc}}{\omega}\right)^2\right] = 0 \tag{2.32}$$

In the following, (2.32) is analyzed in high-, intermediate-, and low-frequency domains; in each frequency domain, (2.32) can be simplified accordingly in the analysis.

2.5.2.1 *High-frequency domain, $\omega > |\Omega_e|$*

The ion terms in (2.32) can be neglected. The resulting equation leads to

(1) $1 - \omega_{pe}^2/\omega(\omega - |\Omega_e|) - (kc/\omega)^2 = 0$ (RH Circular Polarization)

It is associated with the right-hand (RH) circularly polarized EM wave; the dispersion relation is derived to be

$$k_R(\omega) = \left[1 - \frac{\omega_{pe}^2}{\omega(\omega - |\Omega_e|)}\right]^{1/2} (\omega/c) \qquad (2.33)$$

This wave is cutoff from propagation with $\omega \le \omega_{cr} = |\Omega_e|/2 + (\omega_{pe}^2 + \Omega_e^2/4)^{1/2}$.

(2) $1 - \omega_{pe}^2/\omega(\omega + |\Omega_{e1}) - (kc/\omega)^2 = 0$ (LH Circular Polarization)

It is for the left-hand (LH) circularly polarized EM wave, which is governed by the dispersion relation

$$k_L(\omega) = \left[1 - \frac{\omega_{pe}^2}{\omega(\omega + |\Omega_e|)}\right]^{1/2} \left(\frac{\omega}{c}\right) \qquad (2.34)$$

The cutoff frequency is obtained to be $\omega_{c\ell} = -|\Omega_e|/2 + (\omega_{pe}^2 + \Omega_e^2/4)^{1/2}$.

(3) $1 - \omega_{pe}^2/\omega^2 = 0$

It leads to the ES electron plasma oscillation at $\omega = \omega_{pe}$. The dispersion property of the electron plasma wave is given in (2.22).

2.5.2.2 *Intermediate-frequency domain,* $\Omega_i \ll \omega < |\Omega_e|$ *and* $\omega \ll \omega_{pe}^2/|\Omega_e|$

Again, the ion terms in (2.32) can be neglected; only a branch of RH circularly polarized EM modes, governed by $1 - \omega_{pe}^2/\omega(\omega - |\Omega_e|) - k^2c^2/\omega^2 = 0$, can be found from (2.32). Because $\omega_{pe}^2/\omega(|\Omega_e| - \omega) \gg 1$, the dispersion equation is approximated to be

$$\frac{\omega_{pe}^2}{\omega(|\Omega_e| - \omega)} - \left(\frac{kc}{\omega}\right)^2 = 0$$

In the region that $\omega \ll |\Omega_e|$, it gives

$$\omega(k) = \frac{k^2c^2\Omega_i}{\omega_{pi}^2} \tag{2.35}$$

This is the dispersion relation of the whistler wave, whose phase velocity $v_p = (\Omega_i\omega/\omega_{pi}^2)^{1/2}c$ increases with the frequency. As ω is increased to approach $|\Omega_e|$, $k \to \infty$; electron cyclotron resonance occurs at $\omega_R = |\Omega_e|$.

2.5.2.3 *Low-frequency domain,* $\omega \ll \Omega_i$

Equation (2.32) leads to the two dispersion equations $[1 - \sum_a \omega_{pa}^2/\omega(\omega \pm \Omega_a) - (kc/\omega)^2] = 0$, where $\pm$ correspond to the RH/LH circularly polarized EM waves, respectively.

With the aid of $\omega \ll \Omega_i \ll |\Omega_e|$, $\omega_{pe}^2/|\Omega_e| = \omega_{pi}^2/\Omega_i$, and $(\Omega_i \pm \omega)^{-1} \approx \Omega_i^{-1}(1 \mp \omega/\Omega_i)$, both equations are reduced to the same dispersion equation, $1 + \omega_{pi}^2/\Omega_i^2 - k^2c^2/\omega^2 = 0$, for Alfvén waves.

Both RH/LH circularly polarized Alfvén waves are governed by the same dispersion relation

$$\omega(k) = \frac{kc\Omega_i}{\omega_{pi}} = kv_A \tag{2.36}$$

where $1 + \omega_{pi}^2/\Omega_i^2 \approx \omega_{pi}^2/\Omega_i^2$ is assumed and the Alfvén speed $v_A = c\Omega_i/\omega_{pi} = B_0/(n_0 m_i \mu_0)^{1/2}$; thus, (2.36) is also the dispersion relation of linearly polarized Alfvén wave. It is shown that plasma has a relative dielectric constant $\varepsilon_A = \omega_{pi}^2/\Omega_i^2 \gg 1$ in the Alfvén frequency domain. Alfvén wave propagates at much slow speed and without dispersion.

2.5.3 *Dielectric constant of magneto plasma in low-frequency domain*

Consider the propagation of a low-frequency ($\omega \ll \Omega_i$) linearly polarized EM wave along the magnetic field of the plasma. The background magnetic field suppresses the mobility of charged particles in the direction perpendicular to the magnetic field. The wave electric field $\mathbf{E} = \hat{\mathbf{x}}E$ causes electrons and ions to drift together in y-direction with a drift velocity $\mathbf{V}_E = -\hat{\mathbf{y}}E/B_0$, rather than driving them to oscillate in the x-direction. When $\mathbf{E}$ is DC field, a similar case as that presented in Sec. 1.2.1.1, there is no current is induced by the field. On the other hand, a polarization current in the x-direction will be induced by a time varying $\mathbf{E}$.

The time dependence of $\mathbf{V}_E$ indicates that electrons and ions experience y-directed forces $\mathbf{F}_{e,i} = m_{e,i}\, d\mathbf{V}_E/dt = -\hat{\mathbf{y}}\ (m_{e,i}/B_0)\ \partial E/\partial t$, respectively. These forces induce polarization drifts, $\mathbf{V}_{De,i} = \mathbf{B}_0 \times \mathbf{F}_{e,i}/q_{e,i}B_0^2 = \mp(m_{e,i}/eB_0^2)\partial\mathbf{E}/\partial t$, of electrons and ions, which drift oppositely in the x direction. Thus, a polarization current density is induced to be $\mathbf{J} = n_0e(\mathbf{V}_{Di} - \mathbf{V}_{De}) = (n_0m_i/B_0^2)\ \partial\mathbf{E}/\partial t$. Substitute it into (1.22b), which becomes

$$\nabla \times \mathbf{H} = \mathbf{J}_e + \varepsilon_0\left(1 + \frac{n_0m_i}{\varepsilon_0B_0^2}\right)\frac{\partial\mathbf{E}}{\partial t}$$

It shows that the plasma has a relative dielectric constant $\varepsilon_r = 1 + n_0m_i/\varepsilon_0B_0^2 \sim n_0m_i/\varepsilon_0B_0^2$, which is a constant. Thus, wave is non-dispersive and propagates at the phase velocity $v_p = c/\sqrt{\varepsilon_r} = B_0/(n_0m_i\mu_0)^{1/2} = v_A$.

Presented in Fig. 2.1 are plots of dispersion curves governed by (2.33) to (2.36), for different branches of EM modes propagating along the magnetic field.

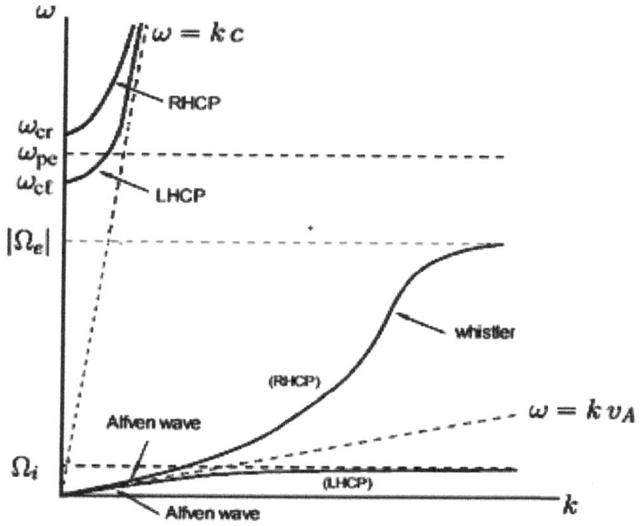

Fig. 2.1. Dispersion relations of circularly polarized EM branches of modes propagating along the magnetic field.

2.5.4 *Propagation transverse to the magnetic field,* $\mathbf{k}_{||} = \mathbf{0}$ *and* $\mathbf{k}_{\perp} = \mathbf{k}$

Equation (2.19) becomes

$$\left[1 - \sum_{a=e,i} \frac{\omega_{pa}^2}{\omega^2} - \left(\frac{kc}{\omega}\right)^2\right] \left\{ \left[1 - \sum_{a=e,i} \frac{\omega_{pa}^2}{\omega(\omega - \Omega_a)}\right] \right.$$
$$\left. \times \left[1 - \sum_{a=e,i} \frac{\omega_{pa}^2}{\omega(\omega - \Omega_a)}\right] - \left(\frac{kc}{\omega}\right)^2 \left[1 - \sum_{a=e,i} \frac{\omega_{pa}^2}{(\omega^2 - \Omega_a^2)}\right] \right\} = 0 \tag{2.37}$$

2.5.4.1 *High-frequency domain,* $\omega > \omega_{pe}$

Equation (2.37) leads to

(1) $1 - \omega_{pe}^2/\omega^2 - (kc/\omega)^2 = 0$ for the ordinary EM mode (O-mode)

It determines a dispersion relation which is the same as (2.31), the dispersion relation of EM mode in unmagnetized plasma. This is because the wave electric field is parallel to the background magnetic field which does not affect the electron motion driven by the wave electric field. This mode is thus called "ordinary" mode.

(2) $[1-\omega_{pe}^2/\omega(\omega-\Omega_e)][1-\omega_{pe}^2/\omega(\omega+\Omega_e)]-(kc/\omega)^2[1-\omega_{pe}^2/(\omega^2-\Omega_e^2)]=0$

It has solutions in two separate frequency bands, $\omega \geq \omega_{cx}$ and $\omega_{c\ell} \leq \omega \leq \omega_{uH}$, being the frequency ranges of the extraordinary mode (X-mode), a hybrid wave of EM nature, and the Z-mode, a hybrid wave of ES nature.

(a) X-mode with $\omega \geq \omega_{cx}$

The dispersion relation is determined to be

$$k_X(\omega) = \left\{\frac{\left[1-\frac{\omega_{pe}^2}{\omega(\omega+|\Omega_e|)}\right]}{\left[1-\frac{\omega_{pe}^2}{(\omega^2-\Omega_e^2)}\right]}\right\}^{1/2}\left[1-\frac{\omega_{pe}^2}{\omega(\omega-|\Omega_e|)}\right]^{1/2}\left(\frac{\omega}{c}\right)$$

$$= \left\{\frac{\left[1-\frac{\omega_{pe}^2}{\omega(\omega+|\Omega_e|)}\right]}{\left[1-\frac{\omega_{pe}^2}{(\omega^2-\Omega_e^2)}\right]}\right\}^{1/2} k_R(\omega) \tag{2.38}$$

Because at $\omega=\omega_{cr}$, $k_R(\omega_{cr})=0=k_X(\omega_{cr})$; ω_{cr} is also the cutoff frequency of the X-mode, i.e., the X-mode cutoff frequency $\omega_{cx}=\omega_{cr}=|\Omega_e|/2+(\omega_{pe}^2+\Omega_e^2/4)^{1/2}$.

(b) Z-mode with $\omega_{c\ell} \leq \omega \leq \omega_{uH}$

The dispersion relation is given by

$$k_Z(\omega) = \left\{\frac{\left[1-\frac{\omega_{pe}^2}{\omega(\omega-|\Omega_e|)}\right]}{\left[1-\frac{\omega_{pe}^2}{(\omega^2-\Omega_e^2)}\right]}\right\}^{1/2}$$

$$\times\left[1-\frac{\omega_{\mathrm{pe}}^2}{\omega(\omega+|\Omega_{\mathrm{e}}|)}\right]^{1/2}\left(\frac{\omega}{\mathrm{c}}\right)$$

$$=\left\{\frac{\left[1-\frac{\omega_{\mathrm{pe}}^2}{\omega(\omega-|\Omega_{\mathrm{e}}|)}\right]}{\left[1-\frac{\omega_{\mathrm{pe}}^2}{(\omega^2-\Omega_{\mathrm{e}}^2)}\right]}\right\}^{1/2}\mathrm{k_L}(\omega) \tag{2.39}$$

Likewise, $\mathrm{k_L}(\omega_{\mathrm{c}\ell}) = 0 = \mathrm{k_Z}(\omega_{\mathrm{c}\ell})$ indicates that $\omega_{\mathrm{c}\ell} = -|\Omega_{\mathrm{e}}|/2 + (\omega_{\mathrm{pe}}^2 + \Omega_{\mathrm{e}}^2/4)^{1/2}$ is also the cutoff frequency of the Z-mode. As the frequency is close to the cutoff frequency, Z-mode is in EM nature. However, as ω is increased to approach the upper hybrid resonance frequency $\omega_{\mathrm{uH}} = (\omega_{\mathrm{pe}}^2 + \Omega_{\mathrm{e}}^2)^{1/2}$, $1 - \omega_{\mathrm{pe}}^2/(\omega^2 - \Omega_{\mathrm{e}}^2) \to 0$, and $\mathrm{k_Z}(\omega) \to \infty$, upper hybrid resonance occurs and Z-mode is in ES nature.

2.5.4.2 *Low-frequency domain,* $\omega^2 \ll (|\Omega_{\mathrm{e}}|\Omega_{\mathrm{i}})^{1/2}$

Equation (2.37) leads to the dispersion relation

$$\omega(\mathrm{k}) = \mathrm{k}\mathrm{c}\Omega_{\mathrm{i}}/\omega_{\mathrm{pi}} = \mathrm{k}\mathrm{v_A} \tag{2.40}$$

This dispersion relation is the same as that presented in (2.36) for parallel propagation Alfvén wave, except it covers a larger frequency band and it is a hybrid mode (as will be shown in Chapter 3) and becomes an EM mode only in the frequency domain $\omega \ll \Omega_{\mathrm{i}}$, where (2.40) is the dispersion relation of the compressional Alfvén wave.

The electric field $\mathbf{E} = \mathbf{E}_{\mathrm{y}}$ of the compressional Alfvén wave is linearly polarized in the direction perpendicular to both the propagation direction ($\hat{\mathbf{k}} = \hat{\mathbf{x}}$) and the background magnetic field $\mathbf{B}_0$, and it forces plasma to drift ($\mathbf{E}\times\mathbf{B}_0/\mathrm{B}_0^2$) into the wave variation, resulting in plasma compression.

Presented in Fig. 2.2 are plots of dispersion curves governed by (2.31) and (2.38) to (2.40), for different branches of EM/hybrid modes propagating perpendicular to the magnetic field. It is noted that the Z-mode dispersion curve and the line $\omega = \mathrm{kc}$ intersect at $\omega = \omega_{\mathrm{pe}}$.

In magneto plasma, wave propagation is direction dependent, the dispersion equation does not have simple analytical solutions.

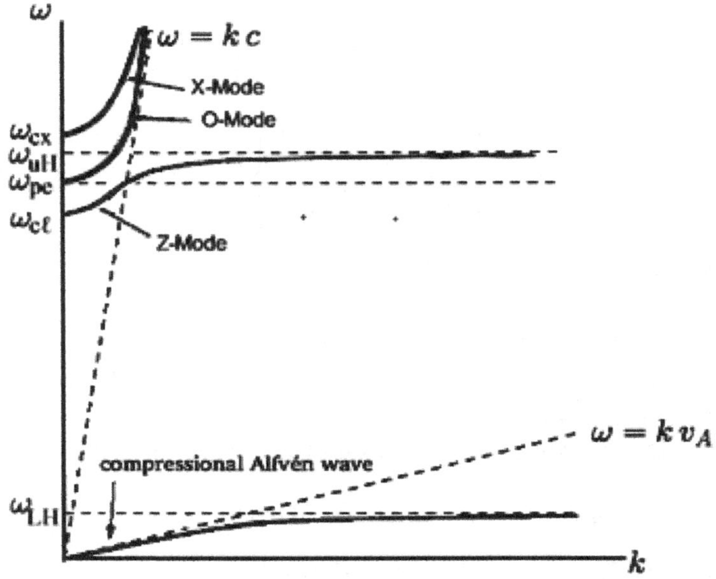

Fig. 2.2. Dispersion relations of EM/hybrid branches of modes propagating perpendicular to the magnetic field.

Normally, two special cases for $\vartheta = 0$ and $\vartheta = \pi/2$ are considered. On the other hand, the dispersion relations of modes in unmagnetized plasma can be derived analytically, it is reduced to three branches of modes, one EM mode governed by (2.31) and two ES modes governed by (2.22) and (2.23). Presented in Fig. 2.3 are dispersion curves of these three branches of modes.

2.6 Wave Dispersion

In a non-dispersive medium, the phase velocity $v_p = \omega/k = c/n$ of a time-harmonic wave is frequency independent, where c and n are the speed of light in free space and the refractive index of the medium; thus, the envelope of a wave packet is preserved. On the other hand, the phase velocity $v_p = c/n(\omega)$ of a time-harmonic wave in a dispersive medium is frequency dependent because of the frequency dependence of the refractive index $n(\omega)$.

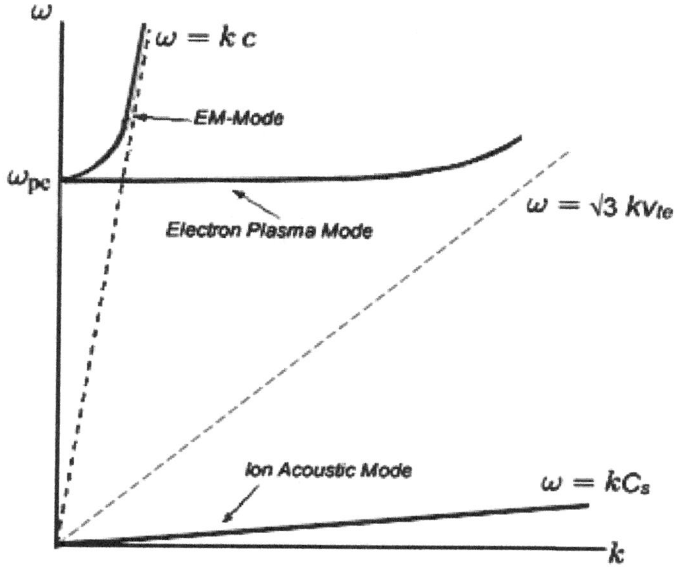

Fig. 2.3. Dispersion relations of EM/ES branches of modes in unmagnetized plasma.

Figures 2.1–2.3 illustrate different branches of EM modes and hybrid modes in magneto and in unmagnetized plasmas. Besides the Alfvén wave branch, all of the other EM dominant branches are dispersive. In other words, except in the Alfvén wave frequency regime in the magnetized case, plasma is a dispersive dielectric medium which disperses waves in the propagation.

This is realized because each frequency component of a wave packet will propagate at a different speed, which leads to phase shifts among the components. It, in turn, causes a spread of the wave packet in the propagation. Hence, the speed of a wave packet is determined by a stationary phase condition, resulting to a group velocity $v_g = \partial\omega/\partial k|_{k=k0}$, where k_0 is the wavenumber of the spectral peak. Wave dispersion is illustrated as follows:

Consider a Gaussian wave packet U(z, t) shown in Fig. 2.4(a) which has a Gaussian spectral distribution

$$A(k) = A_0 \exp[-(k - k_0)^2/L_k] \tag{2.41}$$

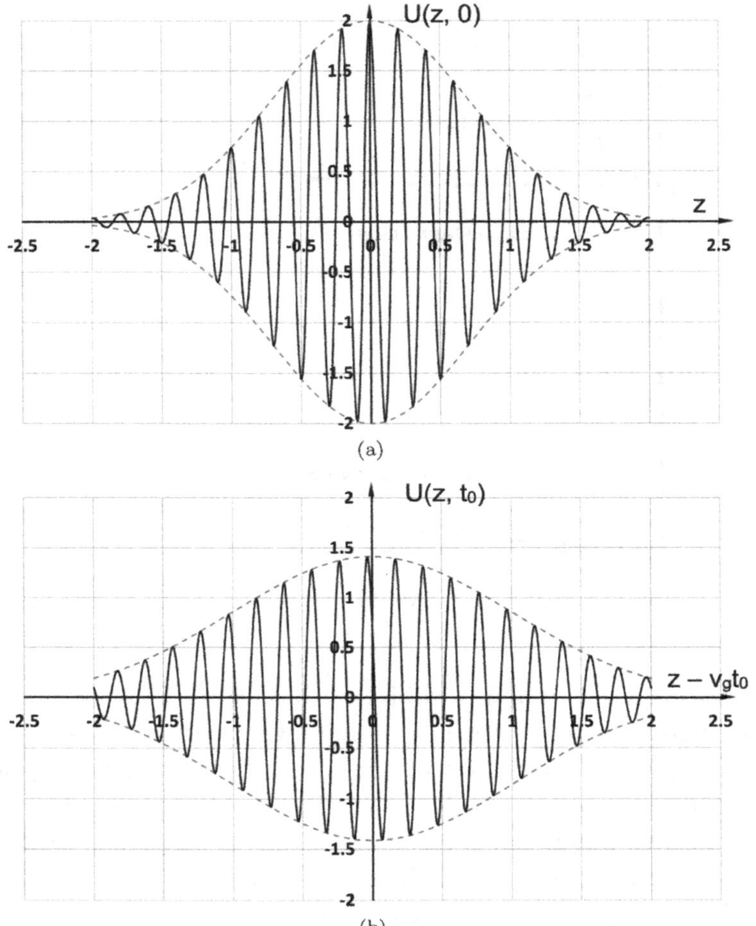

Fig. 2.4. (a) Initial Gaussian wave packet, and (b) dispersed wave packet propagating in a dispersive medium.

Thus,

$$U(z,t) = 1/\sqrt{2\pi}\int_{-\infty}^{\infty} A(k)\exp[i(kz-\omega t)]dk + c.c. \qquad (2.42)$$

With the aid of the expansions

$$k = k_0 + k - k_0$$

and

$$\omega(k) = \omega(k_0) + (k - k_0)\partial\omega/\partial k + (1/2)\,(k - k_0)^2\partial^2\omega/\partial k^2 + \cdots$$

Equation (2.42) becomes

$$\begin{aligned} U(z,t) &\cong (1/\sqrt{2\pi})A_0 \exp[i(k_0 z - \omega_0 t)] \\ &\quad \times \int_{-\infty}^{\infty} \exp[-(k - k_0)^2(1/L_k + ia)] \\ &\quad \times \exp[i(k - k_0)z_1]\; dk + c.c. \\ &= (L_{k1}/2)^{1/2}\; A_0 \exp(-z_1^2 L_{k1}/4) \exp[i(k_0 z - \omega_0 t)] + c.c. \end{aligned} \tag{2.43}$$

where $z_1 = z - v_g\, t$, $L_{k1} = L_k/(1 + ia)$ and $a = (1/2)(\partial^2\omega/\partial k^2)t$.

At $t = 0$, the initial wave packet has the form

$$\begin{aligned} U(z,0) &= (L_k/2)^{1/2}\; A_0 \exp(-z^2 L_k/4) \exp(ik_0 z) + c.c. \\ &= F(z) \exp(ik_0 z) + c.c. \end{aligned} \tag{2.44}$$

where $F(z) = (L_k/2)^{1/2}\; A_0\; \exp(-z^2 L_k/4)$

We now consider two cases,

1. Wave propagation in a non-dispersive medium

In this case, $\partial^2\omega/\partial k^2 = 0$ and $v_g = \omega_0/k_0$, thus,

$$\begin{aligned} U(z,t) &= F(z - v_g t) \exp[ik_0(z - v_g t)] + c.c. \\ &= U(z - v_g t, 0) \end{aligned} \tag{2.45}$$

It shows that the wave packet propagates at a group velocity v_g to preserve its initial wave form.

2. Wave propagation in a dispersive medium, i.e., $\partial^2\omega/\partial k^2 \neq 0$

Let, $\alpha = (1/2)(\partial^2\omega/\partial k^2)$ and $\beta = (\alpha L_k)^2$, and set $L_{k1} = |L_{k1}|e^{-i\theta}$, where $\theta = \tan^{-1}\sqrt{\beta}\,t$ and $|L_{k1}| = L_k/(1 + \beta t^2)^{1/2}$, thus, (43) becomes

$$\begin{aligned} U(z,t) &= (1 + \beta t^2)^{-1/4}\, F[(z - v_g t)/(1 + \beta t^2)^{1/2}] \\ &\quad \times \exp\{i[k_0 z - \omega_0 t + \varphi(z,t)]\} + c.c. \end{aligned} \tag{2.46}$$

where $\varphi(z,t) = \beta(z - v_g t)^2 t/4\alpha(1 + \beta t^2) - (1/2)\tan^{-1}\sqrt{\beta}\,t$.

The wave packet in (46) still propagates at a group velocity v_g; however, its wave form, shown in Fig. 2.4(b), is not preserved in the initial wave form of Fig. 2.4(a) anymore. Besides the space–time-dependent phase shift $\varphi(z, t)$, the envelope of the wave packet is spreading in time in the propagation. While it is spreading, its amplitude decreases due to the conservation of energy. One can show that

$$\int_{-\infty}^{\infty} |U(z,t)|^2 dz = \int_{-\infty}^{\infty} |U(z,0)|^2 dz \tag{2.47}$$

Example:
Determine the spreading rate of a Gaussian wave packet propagating in unmagnetized plasma.

The dispersion relation of the wave is $\omega = (\omega_p^2 + k^2c^2)^{1/2}$; thus, $\partial\omega/\partial k = kc^2/\omega$ and $\partial^2\omega/\partial k^2 = c^2\omega_p^2/\omega^3$, which lead to

$v_g = k_0c^2/\omega_0$, $\omega_0 = (\omega_p^2 + k_0^2c^2)^{1/2}$, $\alpha = c^2\omega_p^2/2\omega_0^3$, and $\beta = (\alpha L_k)^2$.

The width Δz of the wave packet is $\Delta z = [(1+\beta t^2)/L_k]^{1/2}$; thus, the spreading rate $\gamma = \Delta z^{-1} d\Delta z/dt = \beta t/(1 + \beta t^2)$.

A kinetic approach to derive the dielectric tensor of the plasma is introduced in Chapter 3, in which a comprehensive analysis of the plasma modes is given.

Problems

P2.1. Show that the dispersion equation (2.7) for parallel (to the magnetic field) propagating waves in a warm plasma is expressed explicitly to be

$$\varepsilon_3' \left(\varepsilon_1 + \varepsilon_2 - \frac{k^2c^2}{\omega^2}\right)\left(\varepsilon_1 - \varepsilon_2 - \frac{k^2c^2}{\omega^2}\right) = 0$$

where

$$\varepsilon_1 = 1 - \sum_{a=e,i} \frac{\omega_{pa}^2}{\omega^2 - \Omega_a^2}; \quad \varepsilon_2 = -\sum_{a=e,i} \frac{\Omega_a \omega_{pa}^2}{\omega(\omega^2 - \Omega_a^2)};$$

$$\text{and} \quad \varepsilon_3' = 1 - \sum_{a=e,i} \frac{\omega_{pa}^2}{(\omega^2 - \gamma_a k^2 v_{ta}^2)}.$$

P2.2. Show that the dispersion equation (2.7) for transverse (to the magnetic field) propagating waves in a warm plasma is expressed explicitly to be

$$\left(\varepsilon_3 - \frac{k^2c^2}{\omega^2}\right)\left[\varepsilon_1^{'}\left(\varepsilon_1^{''} - \frac{k^2c^2}{\omega^2}\right) - \varepsilon_2^{'2}\right] = 0$$

where

$$\varepsilon_1^{'} = 1 - \sum_{a=e,i} \frac{\omega_{pa}^2}{\omega^2 - \Omega_a^2 - \gamma_a k^2 v_{ta}^2};$$

$$\varepsilon_1^{''} = 1 - \sum_{a=e,i} \frac{\left(1 - \frac{\gamma_a k^2 v_{ta}^2}{\omega^2}\right)\omega_{pa}^2}{\omega^2 - \Omega_a^2 - \gamma_a k^2 v_{ta}^2};$$

$$\varepsilon_2^{'} = -\sum_{a=e,i} \frac{\Omega_a \omega_{pa}^2}{\omega(\omega^2 - \Omega_a^2 - \gamma_a k^2 v_{ta}^2)};$$

$$\text{and} \quad \varepsilon_3 = 1 - \sum_{a=e,i} \frac{\omega_{pa}^2}{\omega^2}.$$

P2.3. Show that $\varepsilon_L(\mathbf{k},\omega) = 1 + i\,\mathbf{k}\cdot\overleftrightarrow{\boldsymbol{\sigma}}(\mathbf{k},\omega)\cdot\mathbf{k}/k^2\omega\varepsilon_0$ in Eq. (2.8) by a test charge approach.
Introduce an external test charge density distribution $\rho_e(k,\omega)\exp[i(\mathbf{k}\cdot\mathbf{r} - \omega t)] + cc$ into plasma, it will perturb the plasma to create charge density fluctuation $\rho(k,\omega)\exp[i(\mathbf{k}\cdot\mathbf{r} - \omega t)] + cc$ and to establish a self-consistent electric field fluctuation $\mathbf{E}(k,\omega)\exp[i(\mathbf{k}\cdot\mathbf{r} - \omega t)] + cc$, which drives a current density fluctuation $\mathbf{J}(k,\omega) = \overleftrightarrow{\boldsymbol{\sigma}}(\mathbf{k},\omega)\cdot\mathbf{E}(k,\omega)$.

(1) Use Eqs. (1.22c) and (1.24) in $\mathbf{k}$–ω space to show that the total charge density fluctuation

$$\rho_{tot}(k,\omega) = \rho_e(k,\omega) + \rho(k,\omega) = \frac{\rho_e(k,\omega)}{\varepsilon_L(\mathbf{k},\omega)}$$

(2) Use Eqs. (1.22c) and (1.23) in $\mathbf{k}$–ω space and the relationship obtained in (1) to show that

$$\varepsilon_L(\mathbf{k},\omega) = 1 + i\frac{\mathbf{k}\cdot\overleftrightarrow{\boldsymbol{\sigma}}(\mathbf{k},\omega)\cdot\mathbf{k}}{k^2\omega\varepsilon_0}$$

P2.4. In an isotropic plasma, the permittivity $\varepsilon = \varepsilon_0(1+\chi_e)$, where χ_e is the electron susceptibility. Expand $n_e = n_0\, e^{e\phi/T}$ for $|e\phi/T| \ll 1$ and calculate electron susceptibility χ_e. Show that $\chi_e \approx 1/k^2\lambda_{De}^2$, where the electron Debye length $\lambda_{De} = (\varepsilon_0 T/n_0 e^2)^{1/2}$.

P2.5. Ion acoustic wave, similar to the sound wave in neutral gas, is a low-frequency pressure wave. Plasma can be considered as one fluid, electrons move together with ion perturbation to keep quasi-neutrality, i.e., $\delta n_e \sim \delta n_i = \delta n$ and $\mathbf{v}_e \sim \mathbf{v}_i = \mathbf{v}$. Consider unmagnetized case,

(1) Use (1.21) to obtain a one-fluid momentum equation (note: $m_i \gg m_e$), which shows that the fluid is driven by the pressure force.

(2) With the aid of the continuity equation (1.20) and neglecting the convective term in the one fluid equation, show that the density perturbation δn is governed by the wave equation

$$(\partial_t^2 - C_s^2\nabla^2)\delta n = 0$$

which leads to the dispersion relation $\omega = kC_s$ of the ion acoustic wave.

P2.6. Upper hybrid resonance oscillation. Add a dc magnetic field $\mathbf{B}_0 = \hat{\mathbf{z}}\, B_0$, which is out of the board, to the slab model presented in Fig. 1.1, and show that Eq. (1.7) is modified to be

$$\frac{d^2\xi_x}{dt^2} = -eE/m_e - |\Omega_e|\frac{d\,\xi_y}{dt}$$
$$= -(\omega_{pe}^2 + \Omega_e^2)\xi_x = -\omega_{uH}^2\,\xi_x$$

P2.7. Lower hybrid resonance. Because of large mass difference, electrons and ions have different oscillation amplitudes in the same AC electric field. To maintain quasi-neutrality for the ES modes in the intermediate- and low-frequency domains, electrons and ions move together in the compressional motion, yet their shear motion can separate considerably. Consider plasma oscillation in a lower hybrid wave field $\mathbf{E} = \hat{\mathbf{x}}\,E(x,t)$, $\Delta x_{e,i}$ represent the compressional displacements of the electron and ion driven by the wave field, thus $d\Delta x_e/dt \cong d\Delta x_i/dt$ to maintain quasi-neutrality. The momentum equation (1.21) in the cold plasma case, can be written as

$$m_a \frac{d^2\Delta \mathbf{r}_a}{dt^2} = q_a \left(\mathbf{E} + \frac{d\Delta \mathbf{r}_a}{dt} \times \mathbf{B}_0 \right)$$

where the subscript "a" $=$ e, i; $\Delta \mathbf{r}_a = \hat{\mathbf{x}}\,\Delta x_a + \hat{\mathbf{y}}\,\Delta y_a; \mathbf{B}_0 = \hat{\mathbf{z}}\,B_0$.

(1) Show that the compressional displacement $\Delta x = \Delta x_i \cong \Delta x_e$ is governed by the equation

$$\frac{d^2\Delta x}{dt^2} + |\Omega_e|\Omega_i \Delta x = 0$$

(2) Show that the velocity ratio of the shear motion is given by

$$|(d\Delta y_e/dt)/(d\Delta y_i/dt)| \sim m_i/m_e.$$

P2.8. Faraday rotation: the polarization (the direction of the wave electric field) of a linearly polarized EM wave rotates around the magnetic field while propagating in plasma along the magnetic field. Use Eqs. (2.30) and (2.31) to show that the relationship between the rotation angle θ and the magnetic field, i.e., $|\Omega_e|$ and the propagation distance z_0 in plasma is given by

$$\theta = [\omega\omega_{pe}^2/(\omega^2 - \Omega_e^2)k_a c^2]\ |\Omega_e| z_0$$

where $k_a = (k_R + k_L)/2$.

[Hint: Assume that the wave enters plasma at $z = 0$ with a wave field $\mathbf{E}(0,t) = \hat{\mathbf{x}}\,E_0 \cos\omega t$. In plasma, it decomposes into two waves, one RHCP and one LHCP wave; thus, $\mathbf{E}(z,t) = (E_0/2)\{[\hat{\mathbf{x}}\cos(k_R z - \omega t) - \hat{\mathbf{y}}\sin(k_R z - \omega t)] + [\hat{\mathbf{x}}\cos(k_L z -$

$\omega t) + \hat{\mathbf{y}} \sin(k_L z - \omega t)]\}$ while propagating in plasma along the magnetic field]

P2.9. Consider the propagation of a RHCP EM wave, with frequency $\omega \gg \Omega_i$, along the magnetic field.

(1) With the aid of (2.30), show that the group velocity of the wave is given by

$$v_{gR} = \frac{\partial \omega}{\partial k} = \frac{2c\omega^{1/2}(\omega - |\Omega_e|)^{\frac{3}{2}}(\omega^2 - |\Omega_e|\omega - \omega_{pe}^2)^{1/2}}{[2\omega(\omega - |\Omega_e|)^2 + |\Omega_e|\omega_{pe}^2]}$$

(2) Show that the group velocity is zero at cutoff frequency $\omega = \omega_{cr}$ and at resonance frequency $\omega = |\Omega_e|$.

P2.10. Energy and power density of EM wave. The time average energy density W of a time harmonic TEM traveling wave in non-dispersive dielectric media is given by $W = W_E + W_M = (1/2)\,\varepsilon_0\varepsilon_r\langle E^2\rangle + (1/2)\,\mu_0\langle H^2\rangle = \varepsilon_0\varepsilon_r\langle E^2\rangle$, where $H = E/\eta$ determined by (1.22a), $\eta = \eta_0/\sqrt{\varepsilon_r}$ and $\eta_0 = (\mu_0/\varepsilon_0)^{1/2}$ is the intrinsic impedance of the free space; the time average power density P is given by the magnitude of the Poynting vector $P = \langle|\mathbf{E} \times \mathbf{H}|\rangle = \langle E^2\rangle/\eta$. In a dispersion medium, $\varepsilon_r = \varepsilon_r(\omega)$, the energy density is modified to be $W = (1/2)\,\varepsilon_0[\partial(\omega\varepsilon_r)/\partial\omega]\langle E^2\rangle + (1/2)\,\mu_0\langle H^2\rangle = (1/2)\,\varepsilon_0\{[\partial(\omega\varepsilon_r)/\partial\omega] + \varepsilon_r\}\,\langle E^2\rangle$.

(1) Show that $P = W\,v_g$ [use the relations $\varepsilon_r(\omega) = (kc/\omega)^2$, $v_g = \partial\omega/\partial k = 1/(\partial k/\partial\omega)$]

(2) Show that $v_g/v_p = 2W_M/(W_E + W_M)$.

P2.11. The impulse produced by a lightning stroke becomes whistler waves to travel along the Earth's magnetic field lines from one hemisphere to the other through the plasma environments of the ionosphere and magnetosphere. Frequencies of terrestrial whistlers are 1 to 30 kHz, with a maximum amplitude usually at 3 to 5 kHz. They undergo dispersion of several kHz due to the increasing frequency dependency of the phase velocity. Thus, whistler waves are perceived as a descending tone which can last for a few seconds.

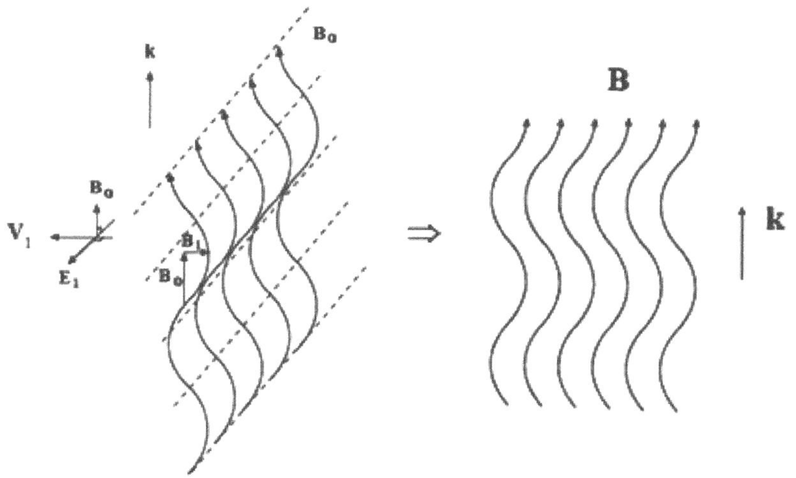

Fig. P2.1. Alfvén wave magnetic field B_1 and total magnetic field $B = (B_0^2 + B_1^2)^{1/2}$.

(1) Show that the group velocity of the whistler wave is twice of the phase velocity.
(2) Show that $W_M \gg W_E$ to justify the relation derived in Problem P2.10(2).

P2.12. The magnetic field line is like a string. A parallel propagating Alfvén wave vibrates the background magnetic field line as shown in Fig. P2.1.

The magnetic field tension provides the restoring force to setup the vibration of the plasma fluid which sticks to the magnetic field line.

Consider an Alfvén wave electric field $\mathbf{E} = \hat{\mathbf{y}}\, E_0 \cos(kz - \omega t)$ in a background magnetic field $\mathbf{B}_0 = \hat{\mathbf{z}}\, B_0$;

(1) Determine the plasma drift velocity $\mathbf{V}_E$
(2) Determine the wave magnetic field $\mathbf{H} = \mathbf{B}/\mu_0$
(3) What is the phase difference between the drift velocity and the total magnetic field intensity $B_t = |\mathbf{B}_0 + \mathbf{B}|$
(4) Because plasma undergoes shear motion, there is no pressure or density fluctuation associated with the wave. Moreover, the displacement current can be neglected (because of low

frequency). Show that $\mathbf{V}_E$ is governed by the one fluid equation

$$\rho_m \frac{\partial \mathbf{V}_E}{\partial t} = \mathbf{J} \times \mathbf{B}_0$$

where the convective terms are neglected, $\rho_m = n(m_i + m_e)$ is the mass density of the fluid, and $\mathbf{J} = e\, n(\mathbf{v}_i - \mathbf{v}_e)$ is the current density driven by the wave field (i.e., in the y-direction).

(5) Apply (1.22a) and (1.22b) to obtain the relationships:

$$\frac{\partial \mathbf{B}}{\partial t} = \nabla \times (\mathbf{V}_E \times \mathbf{B}_0) \text{ and}$$

$$\mathbf{J} \times \mathbf{B}_0 = -\mu_0^{-1} \mathbf{B}_0 \times (\nabla \times \mathbf{B}).$$

Show that the wave magnetic field $\mathbf{B}$ is governed by the wave equation

$$\left(\frac{\partial^2}{\partial t^2} - v_A^2 \nabla^2 \right) \mathbf{B} = 0$$

where $v_A = (B_0^2/\mu_0 \rho_m)^{1/2}$ is the Alfvén speed.

P2.13. The index of refraction $n(\omega)$ of a dielectric medium is defined to be $n(\omega) = kc/\omega = c/v_p$. Show that (1) the group velocity of a wave packet is $v_g = c/[n(\omega) + \omega dn/d\omega]$ and (2) $v_g v_p = c^2$ in unmagnetized plasma.

P2.14. Compare the spreading rates of RHCP and LHCP Gaussian wave packets propagating along the magnetic field of plasma.

Chapter 3

Kinetic-Derivation and Analysis of the Dielectric Tensor

A homogeneous and stationary plasma in uniform dc magnetic field $B_0\hat{\mathbf{z}}$ is considered; it has a phase space density distribution $f_{a0}(\mathbf{v})$ in each species "a". In the presence of an electric field disturbance $\delta\mathbf{E} = \mathbf{E}_1 \exp[i(\mathbf{k} \cdot \mathbf{r} - \omega \mathbf{t})]$, the linearized Vlasov equation (1.18b) is expressed explicitly to be

$$\begin{aligned} &\frac{\partial \delta f_{a1}}{\partial t} + \mathbf{v} \cdot \nabla \delta f_{a1} + \Omega_a \mathbf{v} \times \hat{z} \cdot \nabla_v \delta f_{a1} \\ &\quad = -\frac{q_a}{m_a} \mathbf{E}_1 e^{i(k \cdot r - \omega t)} \cdot \nabla_v f_{a0} \end{aligned} \tag{3.1}$$

where $\Omega_a = q_a B_0/m_a$. There are two ways to describe the dynamics of the perturbed distribution function $\delta f_{a1}(r, v, t)$, Lagrangian and Eulerian specifications. The variables (r, v, t) in the Eulerian specification are independent of one another. In the Lagrangian specification, on the other hand, only t is considered to be the independent variable, both $\mathbf{r}$ and $\mathbf{v}$ are dependent variables which are functions of t, i.e., $\mathbf{r} = \mathbf{r}(t)$ and $\mathbf{v} = \mathbf{v}(t)$; thus, the unperturbed trajectory can be adopted for $\mathbf{r}(t)$ and $\mathbf{v}(t)$ as the zeroth-order functions in the linear analysis. In the magnetized case, it is better to adopt the Lagrangian specification, which follows the phase space fluid element to describe the density variation from the initial state.

3.1 Integration of Linearized Vlasov Equation

The unperturbed trajectory equations of the Lagrangian coordinates are introduced to be

$$\frac{d\mathbf{r}}{dt} = \mathbf{v} \tag{3.2a}$$

and

$$\frac{d\mathbf{v}}{dt} = \Omega_a \mathbf{v} \times \hat{\mathbf{z}} \tag{3.2b}$$

Thus, along the unperturbed trajectory defined by (3.2), (3.1) becomes

$$\frac{d\delta f_{a1}}{dt} = -\frac{q_a}{m_a}\mathbf{E}_1 e^{i(\mathbf{k}\cdot\mathbf{r}-\omega t)} \cdot \nabla_v f_{a0} \tag{3.3}$$

(3.3) is then integrated to be

$$\begin{aligned}\delta f_{a1}(\mathbf{r}, \mathbf{v}, t) &= F_{a1}(\mathbf{v}) e^{i(\mathbf{k}\cdot\mathbf{r}-\omega \mathbf{t})} \\ &= -\frac{q_a}{m_a}\int_{-\infty}^{t} \nabla_{v'} f_{a0}(\mathbf{v}') \cdot \mathbf{E}_1 e^{i(\mathbf{k}\cdot\mathbf{r}'-\omega t')} dt'\end{aligned} \tag{3.4a}$$

where $\mathbf{r}' = \mathbf{r}(t')$ and $\mathbf{v}' = \mathbf{v}(t')$. (3.4a) is reduced to

$$F_{a1}(\mathbf{v}) = -\frac{q_a}{m_a}\int_{-\infty}^{t} \nabla_{v'} f_{a0}(\mathbf{v}') \cdot \mathbf{E}_1 e^{-i[\mathbf{k}\cdot(\mathbf{r}-\mathbf{r}')-\omega(\mathbf{t}-\mathbf{t}')]} dt' \tag{3.4b}$$

We now solve (3.2) for $\mathbf{r}'$ and $\mathbf{v}'$ in terms of $\mathbf{r}$ and $\mathbf{v}$, it yields

$$\mathbf{v}' = \overleftrightarrow{\mathbf{M}}(t'-t) \cdot \mathbf{v} \quad \text{and} \quad \mathbf{r}' = \mathbf{r} + \Omega_a^{-1}\overleftrightarrow{\mathbf{N}}(t'-t) \cdot \mathbf{v} \tag{3.5}$$

where

$$\overleftrightarrow{\mathbf{M}}(t) = \cos\Omega_a t(\hat{\mathbf{x}}\hat{\mathbf{x}} + \hat{\mathbf{y}}\hat{\mathbf{y}}) + \sin\Omega_a t(\hat{\mathbf{x}}\hat{\mathbf{y}} - \hat{\mathbf{y}}\hat{\mathbf{x}}) + \hat{\mathbf{z}}\hat{\mathbf{z}} \tag{3.6a}$$

and

$$\begin{aligned}\overleftrightarrow{\mathbf{N}}(t) &= \sin\Omega_a t(\hat{\mathbf{x}}\hat{\mathbf{x}} + \hat{\mathbf{y}}\hat{\mathbf{y}}) + (1 - \cos\Omega_a t)(\hat{\mathbf{x}}\hat{\mathbf{y}} - \hat{\mathbf{y}}\hat{\mathbf{x}}) \\ &\quad + \Omega_a t\hat{\mathbf{z}}\hat{\mathbf{z}}\end{aligned} \tag{3.6b}$$

Hence, $\mathbf{r} - \mathbf{r}' = -\Omega_a^{-1}\overleftrightarrow{\mathbf{N}}(-\tau) \cdot \mathbf{v}$, where $\tau = t - t'$.

3.1.1 *Integral form of the dielectric tensor*

The induced current density $\delta\mathbf{J} = \mathbf{J}_1 e^{i(\mathbf{k}\cdot\mathbf{r}-\omega t)}$ is calculated through $\delta f_{a1}(\mathbf{r}, \mathbf{v}, t)$; it gives

$$\begin{aligned}
\mathbf{J}_1 &= \sum_{a=e,i} q_a \int \mathbf{v} F_{a1}(\mathbf{v}) d\mathbf{v} \\
&= -\sum_{a=e,i} \frac{q_a^2}{m_a} \int_0^{2\pi} d\theta \int_0^{\infty} \mathbf{v}_\perp d\mathbf{v}_\perp \int_{-\infty}^{\infty} d\mathbf{v}_\parallel \\
&\quad \times \int_0^{\infty} d\tau \mathbf{v} \nabla_{v'} f_{a0}(\mathbf{v}') \cdot \mathbf{E}_1 e^{-i\Phi(\tau,\mathbf{v})} \\
&= \overleftrightarrow{\sigma}(\mathbf{k}, \omega) \cdot \mathbf{E}_1
\end{aligned} \tag{3.7}$$

where

$$\Phi(\tau, \mathbf{v}) = \mathbf{k} \cdot (\mathbf{r} - \mathbf{r}') - \omega(t - t') = -\left[\frac{1}{\Omega_a} \mathbf{k} \cdot \overleftrightarrow{\mathbf{N}}(-\tau) \cdot \mathbf{v} + \omega\tau\right] \tag{3.8}$$

With the aid of (2.5) and (3.7), the dielectric tensor expressed in the integrals of the plasma distribution functions is obtained to be

$$\begin{aligned}
\overleftrightarrow{\varepsilon}(\mathbf{k}, \omega) = \overleftrightarrow{\mathbf{I}} - i \sum_{a=e,i} \frac{\omega_{pa}^2}{n_0 \omega} \int_0^{2\pi} d\theta \int_0^{\infty} \mathbf{v}_\perp d\mathbf{v}_\perp \int_{-\infty}^{\infty} d\mathbf{v}_\parallel \\
\times \int_0^{\infty} d\tau \mathbf{v} \nabla_{v'} f_{a0}(\mathbf{v}') e^{-i\Phi(\tau,\mathbf{v})}
\end{aligned} \tag{3.9}$$

where n_0 is the background plasma density and the plasma frequency $\omega_{pa} = (n_0 q_a^2 / m_a \varepsilon_0)^{1/2}$; with the aid of (3.5) and (3.6), $\nabla_{v'} f_{a0}(\mathbf{v}')$ can be expressed to be dependent on $(\tau, \mathbf{v})$.

3.2 Cylindrical Coordinate Integration with the Aid of Bessel Function Identities

In carrying out the integrations in (3.9), $\mathbf{k} = \hat{\mathbf{x}} k_\perp + \hat{\mathbf{z}} k_\parallel$ is set without losing the generality and $\nabla_{v'} f_{a0}(\mathbf{v}')$ is expressed in terms of variables $\mathbf{v}$ and τ. We first write

$$\mathbf{v} = (\hat{\mathbf{x}}\cos\theta + \hat{\mathbf{y}}\sin\theta)\mathrm{v}_\perp + \hat{\mathbf{z}}\mathrm{v}_\parallel \tag{3.10a}$$

With the aid of (3.5), we have

$$\mathbf{v}'_\mathrm{x} = \mathbf{v}_\perp\cos(\Omega_\mathrm{a}\tau + \theta), \quad \mathbf{v}'_\mathrm{y} = \mathbf{v}_\perp\sin(\Omega_\mathrm{a}\tau + \theta) \tag{3.10b}$$

and

$$\mathbf{v}'_\mathrm{z} = \mathbf{v}'_\parallel = \mathbf{v}_\parallel$$

as well as

$$\begin{aligned}\nabla_{\mathrm{v}'} &= [\hat{\mathbf{x}}\cos(\Omega_\mathrm{a}\tau + \theta) + \hat{\mathbf{y}}\sin(\Omega_\mathrm{a}\tau + \theta)]\frac{\partial}{\partial\mathbf{v}_\perp}\\ &\quad -[\hat{\mathbf{x}}\sin(\Omega_\mathrm{a}\tau + \theta) - \hat{\mathbf{y}}\cos(\Omega_\mathrm{a}\tau + \theta)]\frac{1}{\mathbf{v}_\perp}\frac{\partial}{\partial\theta} + \hat{\mathbf{z}}\frac{\partial}{\partial\mathbf{v}_\parallel}\end{aligned} \tag{3.10c}$$

Because $\mathrm{f_{a0}}(\mathbf{v}')$ given in (1.19) is isotropic, $\mathrm{f_{a0}}(\mathbf{v}') = \mathrm{f_{a0}}(\mathrm{v}_\perp, \mathrm{v}_\parallel)$ which leads to

$$\nabla_{\mathrm{v}'}\mathrm{f_{a0}}(\mathbf{v}') = [\hat{\mathbf{x}}\cos(\Omega_\mathrm{a}\tau + \theta) + \hat{\mathbf{y}}\sin(\Omega_\mathrm{a}\tau + \theta)]\frac{\partial \mathrm{f_{a0}}}{\partial \mathrm{v}_\perp} + \hat{\mathbf{z}}\frac{\partial \mathrm{f_{a0}}}{\partial \mathrm{v}_\parallel} \tag{3.11}$$

With the aid of (3.6) and (3.10a), (3.7) becomes

$$\Phi(\tau, \mathbf{v}) = \frac{\mathrm{k}_\perp \mathrm{v}_\perp}{\Omega_\mathrm{a}}[\sin(\theta + \Omega_\mathrm{a}\tau) - \sin\theta] + (\mathrm{k}_\parallel \mathrm{v}_\parallel - \omega)\tau \tag{3.12a}$$

3.2.1 *Expansions in terms of Bessel functions*

Employ the expansion $\mathrm{e}^{-i\alpha\sin\Psi} = \sum_{\mathrm{n}=-\infty}^{\infty}\mathrm{J_n}(\alpha)\mathrm{e}^{-in\Psi}$ through the nth order Bessel functions $\mathrm{J_n}(\alpha)$ and introduce $\varsigma_\mathrm{a} = \mathrm{k}_\perp\mathrm{v}_\perp/\Omega_\mathrm{a}$, then we have

$$\begin{aligned}\mathrm{e}^{-i\Phi(\tau,\mathbf{v})} &= \sum_{\mathrm{n}=-\infty}^{\infty}\sum_{\mathrm{n}'=-\infty}^{\infty}\mathrm{J_n}(\varsigma_\mathrm{a})\mathrm{J_{n'}}(\varsigma_\mathrm{a})\exp\{-\mathrm{i}[\mathrm{n}(\Omega_\mathrm{a}\tau + \theta) - \mathrm{n}'\theta]\}\\ &\quad \times \exp[-\mathrm{i}(\mathrm{k}_\parallel \mathrm{v}_\parallel - \omega)\tau]\end{aligned} \tag{3.12b}$$

Before carrying out the time integration in (3.9), the following two relations are applied:

$$\left\{ [\hat{\mathbf{x}}\cos(\Omega_a\tau+\theta)+\hat{\mathbf{y}}\sin(\Omega_a\tau+\theta)]\frac{\partial f_{a0}}{\partial v_\perp}+\hat{\mathbf{z}}\frac{\partial f_{a0}}{\partial v_\parallel}\right\}$$
$$\times\sum_{n=-\infty}^{\infty} J_n(\varsigma_a)\exp[-in(\Omega_a\tau+\theta)]$$
$$=\sum_{n=-\infty}^{\infty}\left\{\left[\frac{n}{\varsigma_a}J_n(\varsigma_a)\hat{\mathbf{x}}+iJ_n'(\varsigma_a)\hat{\mathbf{y}}\right]\frac{\partial f_{a0}}{\partial v_\perp}+\hat{\mathbf{z}}\,J_n(\varsigma_a)\frac{\partial f_{a0}}{\partial v_\parallel}\right\}$$
$$\times\exp[-in(\Omega_a\tau+\theta)] \tag{3.13}$$

and

$$(\hat{\mathbf{x}}v_\perp\cos\theta+\hat{\mathbf{y}}v_\perp\sin\theta+\hat{\mathbf{z}}v_\parallel)\sum_{n'=-\infty}^{\infty}J_{n'}(\varsigma_a)\exp(in'\theta)$$
$$=\sum_{n'=-\infty}^{\infty}\left\{\left[\frac{n'}{\varsigma_a}J_{n'}(\varsigma_a)\hat{\mathbf{x}}-iJ_{n'}'(\varsigma_a)\hat{\mathbf{y}}\right]v_\perp+\hat{\mathbf{z}}J_{n'}(\varsigma_a)v_\parallel\right\}\exp(in'\theta) \tag{3.14}$$

where the similar procedures of

$$\cos\Psi\sum_{n=-\infty}^{\infty}J_n(y)\exp(in\Psi)=1/2\sum_{n=-\infty}^{\infty}J_n(y)[e^{i(n+1)\Psi}+e^{i(n-1)\Psi}]$$
$$=1/2\sum_{n=-\infty}^{\infty}[J_{n-1}(y)+J_{n+1}(y)]e^{in\Psi}$$

and

$$\sin\Psi\sum_{n=-\infty}^{\infty}J_n(y)\exp(in\Psi)=-\frac{i}{2}\sum_{n=-\infty}^{\infty}J_n(y)[e^{i(n+1)\Psi}-e^{i(n-1)\Psi}]$$
$$=-\frac{i}{2}\sum_{n=-\infty}^{\infty}[J_{n-1}(y)-J_{n+1}(y)]e^{in\Psi}$$

are adopted; the Bessel function identities of

$$J_{n-1}(y) + J_{n+1}(y) = (2n/y)J_n(y)$$

and

$$J_{n-1}(y) - J_{n+1}(y) = 2J'_n(y)$$

are used; and $J'_n(y) = dJ_n(y)/dy$.

3.2.2 *Identities used*

With the aid of (3.10a), (3.11), (3.12b), (3.13), and (3.14) and the condition:

$$\int_0^{2\pi} d\theta \exp[-i(n-n')\theta] = 2\pi\delta_{nn'}$$

where $\delta_{nn'}$ is the Kronecker delta function, (i.e., $\delta_{nn'} = 1$ if $n' = n$, and $\delta_{nn'} = 0$, if $n' \neq n$), the time integration in (3.9) can be carried out straightforwardly; so we obtain

$$\int_0^{2\pi} d\theta \int_0^{\infty} d\tau \mathbf{v}\nabla_{v'} f_{a0}(\mathbf{v}')e^{-i\Phi(\tau,\mathbf{v})}$$

$$= -i2\pi \sum_{n=-\infty}^{\infty} \left\{ \left[\frac{n}{\varsigma_a} J_n(\varsigma_a)\hat{\mathbf{x}} - iJ'_n(\varsigma_a)\hat{\mathbf{y}} \right] v_\perp + \hat{\mathbf{z}} J_n(\varsigma_a) v_\parallel \right\}$$

$$\times \left\{ \left[\frac{n}{\varsigma_a} J_n(\varsigma_a)\hat{\mathbf{x}} + iJ'_n(\varsigma_a)\hat{\mathbf{y}} \right] \frac{\partial f_{a0}}{\partial v_\perp} + \hat{\mathbf{z}} J_n(\varsigma_a) \frac{\partial f_{a0}}{\partial v_\parallel} \right\} \frac{1}{(k_\parallel v_\parallel + n\Omega_a - \omega)}$$

$$= -i2\pi \sum_{n=-\infty}^{\infty} \left\{ \left[\left(\frac{n\Omega_a}{k_\perp} \right)^2 J_n^2 \hat{\mathbf{x}}\hat{\mathbf{x}} + iv_\perp \right. \right.$$

$$\left. \times \left(\frac{n\Omega_a}{k_\perp} \right) J_n J'_n (\hat{\mathbf{x}}\hat{\mathbf{y}} - \hat{\mathbf{y}}\hat{\mathbf{x}}) + v_\perp^2 J'^2_n \hat{\mathbf{y}}\hat{\mathbf{y}} \right] \frac{1}{v_\perp} \frac{\partial f_{a0}}{\partial v_\perp}$$

$$+ v_\parallel \left(\frac{n\Omega_a}{k_\perp} \right) J_n^2 \left(\frac{1}{v_\parallel} \frac{\partial f_{a0}}{\partial v_\parallel} \hat{\mathbf{x}}\hat{\mathbf{z}} + \frac{1}{v_\perp} \frac{\partial f_{a0}}{\partial v_\perp} \hat{\mathbf{z}}\hat{\mathbf{x}} \right)$$

$$- \mathrm{i}v_{\|}v_{\perp}J_nJ_n'\left(\frac{1}{v_{\|}}\frac{\partial f_{a0}}{\partial v_{\|}}\hat{\mathbf{y}}\hat{\mathbf{z}} - \frac{1}{v_{\perp}}\frac{\partial f_{a0}}{\partial v_{\perp}}\hat{\mathbf{z}}\hat{\mathbf{y}}\right)$$

$$+ (v_{\|}^2J_n^2)\frac{1}{v_{\|}}\frac{\partial f_{a0}}{\partial v_{\|}}\hat{\mathbf{z}}\hat{\mathbf{z}}\Bigg\}\frac{1}{k_{\|}v_{\|} + n\Omega_a - \omega} \tag{3.15}$$

The terms on the right-hand side (RHS) of (3.15) can be regrouped to simplify the expression. This is done with the aid of the following relations. First, the Maxwellian distribution given in (1.19) indicates that $v_{\perp}^{-1}\partial f_{a0}/\partial v_{\perp} = v_{\|}^{-1}\partial f_{a0}/\partial v_{\|}$, which is then applied to obtain the relations:

$$\begin{aligned}(k_{\|}v_{\|} + n\Omega_a)\frac{1}{v_{\perp}}\frac{\partial f_{a0}}{\partial v_{\perp}} &= \left[\frac{n\Omega_a}{v_{\perp}}\frac{\partial f_{a0}}{\partial v_{\perp}} + k_{\|}\frac{\partial f_{a0}}{\partial v_{\|}}\right]\\ &= (k_{\|}v_{\|} + n\Omega_a)\frac{1}{v_{\|}}\frac{\partial f_{a0}}{\partial v_{\|}};\end{aligned}$$

second, employ the Bessel function identities:

$$\sum_{n=-\infty}^{\infty} J_n^2(\varsigma_a) = 1, \quad \sum_{n=-\infty}^{\infty} [nJ_n(\varsigma_a)]^2 = \frac{\varsigma_a^2}{2}, \quad \sum_{n=-\infty}^{\infty} J_n'^2(\varsigma_a) = 1,$$

$$\sum_{n=-\infty}^{\infty} nJ_n^2(\varsigma_a) = 0, \quad \sum_{n=-\infty}^{\infty} J_n(\varsigma_a)J_n'(\varsigma_a) = 0, \quad \sum_{n=-\infty}^{\infty} nJ_n(\varsigma_a)J_n'(\varsigma_a) = 0;$$

and third, express

$$\frac{\omega}{(k_{\|}v_{\|} + n\Omega_a - \omega)} = \frac{(k_{\|}v_{\|} + n\Omega_a)}{(k_{\|}v_{\|} + n\Omega_a - \omega)} - 1.$$

3.2.3 *Dielectric tensor in Bessel function integrations*

Thus, (3.15) becomes

$$\int_0^{2\pi} d\theta \int_0^{\infty} d\tau \mathbf{v}\nabla_{\mathbf{v}'}f_{a0}(\mathbf{v}')e^{-\mathrm{i}\Phi(\tau,\mathbf{v})}$$

$$= -\mathrm{i}\frac{2\pi}{\omega}\sum_{n=-\infty}^{\infty}\left\{\left[\left(\frac{n\Omega_a}{k_{\perp}}\right)^2 J_n^2\hat{\mathbf{x}}\hat{\mathbf{x}} + \mathrm{i}v_{\perp}\right.\right.$$

$$\times\left(\frac{n\Omega_a}{k_\perp}\right)J_nJ_n'(\hat{\mathbf{x}}\hat{\mathbf{y}}-\hat{\mathbf{y}}\hat{\mathbf{x}})+v_\perp^2J_n'^2\hat{\mathbf{y}}\hat{\mathbf{y}}\Bigg]$$

$$+v_\parallel\left(\frac{n\Omega_a}{k_\perp}\right)J_n^2(\hat{\mathbf{x}}\hat{\mathbf{z}}+\hat{\mathbf{z}}\hat{\mathbf{x}})-iv_\parallel v_\perp J_nJ_n'(\hat{\mathbf{y}}\hat{\mathbf{z}}-\hat{\mathbf{z}}\hat{\mathbf{y}})+(v_\parallel^2J_n^2)\hat{\mathbf{z}}\hat{\mathbf{z}}\Bigg]$$

$$\times\left[\frac{n\Omega_a}{v_\perp}\frac{\partial f_{a0}}{\partial v_\perp}+k_\parallel\frac{\partial f_{a0}}{\partial v_\parallel}\right]\frac{1}{(k_\parallel v_\parallel+n\Omega_a-\omega)}\Bigg\}$$

$$-i\left(\frac{2\pi}{\omega}\right)\left(\frac{m_a}{2T_a}\right)[(\hat{\mathbf{x}}\hat{\mathbf{x}}+\hat{\mathbf{y}}\hat{\mathbf{y}})v_\perp^2+\hat{\mathbf{z}}\hat{\mathbf{z}}2v_\parallel^2]f_{a0} \tag{3.16}$$

Substitute (3.16) into (3.9), which becomes

$$\overleftrightarrow{\boldsymbol{\varepsilon}}(\mathbf{k},\omega)=(1-\omega_p^2/\omega^2)\overleftrightarrow{\mathbf{I}}-\sum_{a=e,i}\frac{\omega_{pa}^2}{n_0\omega^2}\sum_{n=-\infty}^{\infty}2\pi\int_0^\infty v_\perp dv_\perp\int_{-\infty}^{\infty}dv_\parallel$$

$$\times\left\{\left[\frac{n\Omega_a}{v_\perp}\frac{\partial f_{a0}}{\partial v_\perp}+k_\parallel\frac{\partial f_{a0}}{\partial v_\parallel}\right]\frac{1}{(k_\parallel v_\parallel+n\Omega_a-\omega)}\right\}\overleftrightarrow{\mathbf{K}}_a \tag{3.17}$$

where $\omega_p^2=\sum_{a=e,i}\omega_{pa}^2$, and the tensor

$$\overleftrightarrow{\mathbf{K}}_a=\left[\left(\frac{n\Omega_a}{k_\perp}\right)^2J_n^2\hat{\mathbf{x}}\hat{\mathbf{x}}+iv_\perp\left(\frac{n\Omega_a}{k_\perp}\right)J_nJ_n'(\hat{\mathbf{x}}\hat{\mathbf{y}}-\hat{\mathbf{y}}\hat{\mathbf{x}})+v_\perp^2J_n'^2\hat{\mathbf{y}}\hat{\mathbf{y}}\right]$$

$$+v_\parallel\left(\frac{n\Omega_a}{k_\perp}\right)J_n^2(\hat{\mathbf{x}}\hat{\mathbf{z}}+\hat{\mathbf{z}}\hat{\mathbf{x}})-iv_\parallel v_\perp J_nJ_n'(\hat{\mathbf{y}}\hat{\mathbf{z}}-\hat{\mathbf{z}}\hat{\mathbf{y}})+(v_\parallel^2J_n^2)\hat{\mathbf{z}}\hat{\mathbf{z}}\Bigg] \tag{3.18}$$

where $J_n=J_n(\varsigma_a)$ and $J_n'=J_n'(\varsigma_a)$.

On the RHS of (3.17), the first term is ascribed to the dielectric response function $\varepsilon_r(\omega)=1-\omega_p^2/\omega^2$ of a cold isotropic plasma (i.e., in the case of $T_a=0=B_0$). The second term, given by the summation, measures the effect of background magnetic field, which introduces anisotropy to the plasma and causes gyrations of charged particles, and the thermal effect on the plasma response to the electric disturbance. The thermal pressure of the plasma modifies the plasma collective effect; it enables the plasma disturbances to propagate under the mode conditions (i.e., satisfying a dispersion relation). Anisotropy modifies the propagation characteristics of the plasma modes, both electrostatic and electromagnetic; moreover, it also gives rise hybrid modes. The gyrations of charged particles give rise to resonance phenomena.

3.2.4 *Identities of Bessel function integrations*

We now apply the general result of a Bessel function integration

$$\int_0^\infty \exp(-a^2x^2)J_p(hx)J_p(gx)x\,dx = \frac{1}{2a^2}\exp\left[-\frac{(h^2+g^2)}{4a^2}\right]I_p\left(\frac{hg}{2a^2}\right) = F_p(h,g;a^2) \qquad (3.19)$$

to carry out the following three integrations:

$$\int_0^\infty \exp\left(-\frac{m_a v_\perp^2}{2T_a}\right)J_n^2(\varsigma_a)v_\perp dv_\perp = F_n\left(\frac{k_\perp}{\Omega_a},\frac{k_\perp}{\Omega_a};\frac{m_a}{2T_a}\right) = \frac{T_a}{m_a}\Lambda_n(\beta_a) \qquad (3.20a)$$

$$\int_0^\infty \exp\left(-\frac{m_a v_\perp^2}{2T_a}\right)J_n(\varsigma_a)J_n'(\varsigma_a)v_\perp^2 dv_\perp = \left.\frac{\partial F_n(h,g;\frac{m_a}{2T_a})}{\partial h}\right|_{h=g=k_\perp/\Omega_a} = \left(\frac{T_a}{m_a}\right)^{\frac{3}{2}}\beta_a^{1/2}\Lambda_n'(\beta_a) \qquad (3.20b)$$

and

$$\int_0^\infty \exp\left(-\frac{m_a v_\perp^2}{2T_a}\right)J_n'^2(\varsigma_a)v_\perp^3 dv_\perp = \left.\frac{\partial^2 F_n(h,g;\frac{m_a}{2T_a})}{\partial h\partial g}\right|_{h=g=k_\perp/\Omega_a} = \left(\frac{T_a}{m_a}\right)^2\left[\frac{n^2}{\beta_a}\Lambda_n(\beta_a) - 2\beta_a\Lambda_n'(\beta_a)\right] \qquad (3.20c)$$

where $\beta_a = k_\perp^2 T_a/m_a\Omega_a^2$, $\Lambda_n(\beta_a) = I_n(\beta_a)\exp(-\beta_a)$, $\Lambda'(\beta_a) = d\Lambda_n(\beta_a)/d\beta_a$, and I_n is the modified Bessel function of the first kind.

3.2.5 *Plasma dispersion function and the explicit function of the dielectric tensor*

We next introduce a plasma dispersion function

$$Z(\xi) = \frac{1}{\sqrt{2\pi}} \int_{-\infty}^{\infty} dx \frac{\exp(-\frac{x^2}{2})}{x - \xi} \tag{3.21}$$

With the aid of (3.20) and (3.21), (3.17) becomes

$$\overleftrightarrow{\varepsilon}(\mathbf{k}, \omega) = \overleftrightarrow{I} + \sum_{a=e,i} \frac{\omega_{pa}^2}{\omega^2} \left\{ \sum_{n=-\infty}^{\infty} [\xi_{a0} Z(\xi_{an}) \overleftrightarrow{\Pi}] + \xi_{a0}^2 \hat{\mathbf{z}}\hat{\mathbf{z}} \right\} \tag{3.22}$$

where $\xi_{an} = (\omega - n\Omega_a)/k_{\|} v_{ta}$, $v_{ta} = (T_a/m_a)^{1/2}$, and

$$\begin{aligned}\overleftrightarrow{\Pi} = {} & \frac{n^2}{\beta_a} \Lambda_n(\beta_a) \hat{\mathbf{x}}\hat{\mathbf{x}} + in\Lambda_n'(\beta_a)(\hat{\mathbf{x}}\hat{\mathbf{y}} - \hat{\mathbf{y}}\hat{\mathbf{x}}) \\ & + \left[\frac{n^2}{\beta_a} \Lambda_n(\beta_a) - 2\beta_a \Lambda_n'(\beta_a)\right] \hat{\mathbf{y}}\hat{\mathbf{y}} + \xi_{an} \left(\frac{n}{\sqrt{\beta_a}}\right) \Lambda_n(\beta_a)(\hat{\mathbf{x}}\hat{\mathbf{z}} + \hat{\mathbf{z}}\hat{\mathbf{x}}) \\ & - i\xi_{an} \sqrt{\beta_a} \Lambda_n'(\beta_a)(\hat{\mathbf{y}}\hat{\mathbf{z}} - \hat{\mathbf{z}}\hat{\mathbf{y}}) + \xi_{an}^2 \Lambda_n(\beta_a) \hat{\mathbf{z}}\hat{\mathbf{z}}\end{aligned} \tag{3.23}$$

The plasma dispersion function $Z(\xi)$ can be expanded in the two regions, $|\xi| < 1$ and $|\xi| \gg 1$, as

$$Z(\xi) = i\left(\frac{\pi}{2}\right)^{1/2} \exp\left(-\frac{\xi^2}{2}\right) - \sum_{p=0}^{\infty} (-1)^p \frac{\xi^{2p+1}}{(2p+1)!!} \quad \text{for } |\xi| < 1 \tag{3.24a}$$

and

$$Z(\xi) = i\left(\frac{\pi}{2}\right)^{1/2} \exp\left(-\frac{\xi^2}{2}\right) - \sum_{p=0}^{\infty} \left[\frac{(2p-1)!!}{\xi^{2p+1}}\right] \quad \text{for } |\xi| \gg 1 \tag{3.24b}$$

where the double factorial function $(2p \pm 1)!! = 1 \times 3 \times 5 \cdots \times (2p \pm 1)$ for $2p \pm 1 > 0$, and $(2p \pm 1)!! = 1$ for $2p \pm 1 = 0$ or -1. The imaginary terms in the expanded plasma dispersion functions (3.24a and 3.24b) are attributed to the pole in the integral of (3.21). Physically, this pole represents a resonance interaction between the wave and charge particles, which results in a damping of the wave in thermal plasma, named Landau damping.

3.3 Plasma Dielectric Function of Electrostatic Modes

Set $\hat{\mathbf{k}} = (k_\perp/k)\hat{\mathbf{x}} + (k_\parallel/k)\hat{\mathbf{z}}$ in (2.8); with the aid of (3.22), the dispersion equation characterizing electrostatic modes is derived to be

$$\varepsilon_L(\mathbf{k},\omega) = \hat{\mathbf{k}} \cdot \overleftrightarrow{\varepsilon}(\mathbf{k},\omega) \cdot \hat{\mathbf{k}}$$
$$= 1 + \sum_{a=e,i} \frac{\omega_{pa}^2}{k^2 v_{ta}^2} \left\{ 1 + \sum_{n=-\infty}^{\infty} [\xi_{a0} Z(\xi_{an}) \Lambda_n(\beta_a)] \right\} = 0 \tag{3.25}$$

In the following, (3.25) is analyzed in the unmagnetized and magnetized cases separately.

3.3.1 *Unmagnetized plasma*

In the absence of background magnetic field, the medium is isotropic and $\mathbf{k} = \hat{\mathbf{z}}\, k$ is assumed. Substitute $\beta_a = 0$ and $\xi_{an} = \xi_{a0} = \omega/kv_{ta}$ into (3.25), it becomes

$$1 + \sum_{a=e,i} \frac{\omega_{pa}^2}{k^2 v_{ta}^2} [1 + \xi_{a0} Z(\xi_{a0})] = 0 \tag{3.26}$$

3.3.1.1 *Electron plasma (Langmuir) wave*

We first consider high-frequency oscillation that ion is too heavy to follow it, thus ion terms in (3.26) are neglected. We next use (3.24b) to expand $Z(\xi_{e0})$ for $|\xi_{e0}| \gg 1$. Thus, (3.26) becomes

$$1 - \frac{\omega_{pa}^2}{\omega^2}\left(1 + \frac{3k^2 v_{te}^2}{\omega^2}\right) + i\left(\frac{\pi}{2}\right)^{1/2}\left(\frac{k_{De}^2}{k^2}\right)\left(\frac{\omega}{kv_{te}}\right)\exp\left(-\frac{\omega^2}{2k^2 v_{te}^2}\right) = 0 \tag{3.27}$$

where $k_{De} = \omega_{pe}/v_{te}$ is the electron Debye number. Substitute $\omega = \omega_{ek} + i\gamma_{ek}$ into (3.27), the real frequency ω_{ek} and the damping rate

γ_{ek} of the oscillation are obtained to be

$$\omega_{ek} = (\omega_{pe}^2 + 3k^2 v_{te}^2)^{1/2} \tag{3.28a}$$

and

$$\gamma_{ek} = -\left(\frac{\pi}{8}\right)^{1/2} \left(\frac{k_{De}^2}{k^2}\right) \left(\frac{\omega_{ek}^2}{kv_{te}}\right) \exp\left(-\frac{\omega_{ek}^2}{2k^2 v_{te}^2}\right) \tag{3.28b}$$

Equation (3.28a) is the dispersion relation of the electron plasma (Langmuir) wave, which is illustrated in Sec. 1.1.2 of Chapter 1 and is the same as (2.22) derived via the fluid approach. The Landau damping rate is, however, revealed in the kinetic approach, which includes the electron–wave resonance interaction (i.e., the speed of the electron $v \sim \omega_{ek}/k$, the phase speed of the wave) in the kinetic formulation.

3.3.2 *Landau damping*

The Landau damping rate (3.28b) is proportional to the density ($\propto \exp(-\omega_{ek}^2/2k^2 v_{te}^2)$) of the resonance electrons which move close to the wave phase velocity $v_p = \omega_{ek}/k$. A physical interpretation of the Landau damping is given in the following.

In the moving frame $V = v_0 = \omega_{ek}/k$ of the wave, the wave field becomes a dc field introducing a spatially periodic potential distribution $\phi(x) = \phi_0 \cos kx$. Those electrons moving closely together with the wave can be trapped in the potential well $-e\phi(x)$ as shown in Fig. 3.1, for example, in the region between $x = -\lambda/2$

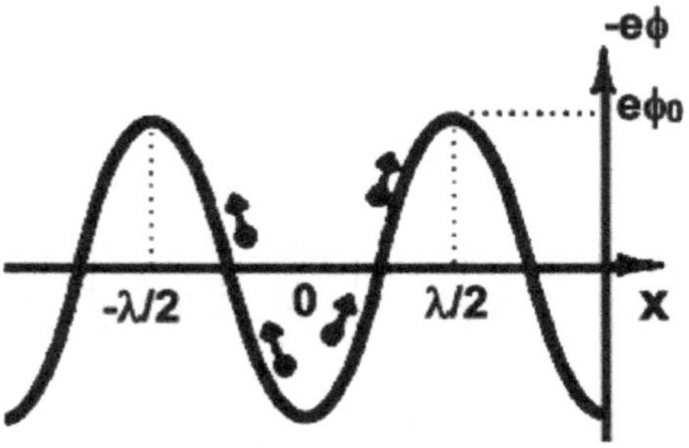

Fig. 3.1. Electrons trapped in the potential well of the Langmuir wave.

and $\lambda/2$, and interact strongly with the wave. Among the trapped electrons, those moving, with velocity $v = v_0 + \Delta v$, faster than the wave will be slow down and those, with $v = v_0 - \Delta v$, slower than the wave will be accelerated. In one half of a bounce period $t_B = 2\pi/k(\pi e\phi_0/4m_e)^{1/2}$, there will be a net energy exchange between the wave and the trapped electrons. The density of slower and faster electrons are approximately $f_{ex}(v_0 - \Delta v_0/2)$ and $f_{ex}(v_0 + \Delta v_0/2)$, where $f_{ex}(v) = n_0(m_e/2\pi T_e)^{1/2}\exp(-m_e v^2/2T_e)$ is the one-dimensional Maxwellian distribution and Δv_0 is the maximum velocity deviation of the trapped electrons from the phase velocity of the wave, i.e., $1/2\ m_e\Delta v_0^2 = (\pi/2)e\phi_0$, giving $\Delta v_0 = (\pi e\phi_0/m_e)^{1/2}$. Thus, the net power transfer from the wave to the electrons per unit spatial volume is

$$\left[f_{ex}\left(v_0 - \frac{\Delta v_0}{2}\right) - f_{ex}\left(v_0 + \frac{\Delta v_0}{2}\right)\right]\Delta v_0\left(\frac{m_e v_0 \Delta v_0/2}{t_B/4}\right)$$
$$= -\frac{2f'_{ex}(v_0)m_e v_0 \Delta v_0^3}{t_B}$$
$$= \frac{1}{2}\varepsilon_0\left(\frac{\pi}{2}\right)^{1/2} k_{De}^2\left(\frac{\omega_{ek}^2}{kv_{te}}\right)\exp\left(-\frac{\omega_{ek}^2}{2k^2v_{te}^2}\right)\phi_0^2$$

The power balance equation is

$$\frac{d[(1/2)\,\varepsilon_0 k^2\phi_0^2]}{dt} = -\frac{1}{2}\varepsilon_0\left(\frac{\pi}{2}\right)^{1/2} k_{De}^2\left(\frac{\omega_{ek}^2}{kv_{te}}\right)\exp\left(-\frac{\omega_{ek}^2}{2k^2v_{te}^2}\right)\phi_0^2$$

It is reduced to

$$\frac{d[\phi_0^2]}{dt} = -\left(\frac{\pi}{2}\right)^{1/2}\left(\frac{k_{De}^2}{k^2}\right)\left(\frac{\omega_{ek}^2}{kv_{te}}\right)\exp\left(-\frac{\omega_{ek}^2}{2k^2v_{te}^2}\right)\phi_0^2 = 2\gamma_{ek}\phi_0^2$$

which shows that the wave damps with a damping rate γ_{ek} given by (3.28b).

3.3.2.1 *Ion acoustic wave*

In the low-frequency domain, $kv_{te} \gg |\omega| \gg kv_{ti}$, both electrons and ions can follow the temporal variation of the wave; using (3.24a) and (3.24b) to expand $Z(\xi_{e0})$ and $Z(\xi_{i0})$, respectively, (3.25) becomes

$$1+\left(\frac{k_{De}^2}{k^2}\right)-\frac{\omega_{pi}^2}{\omega^2}\left(1+\frac{3k^2v_{ti}^2}{\omega^2}\right)+i\left(\frac{\pi}{2}\right)^{1/2}$$
$$\times\left[\left(\frac{k_{De}^2}{k^2}\right)\left(\frac{\omega}{kv_{te}}\right)+\left(\frac{k_{Di}^2}{k^2}\right)\left(\frac{\omega}{kv_{ti}}\right)\exp\left(-\frac{\omega^2}{2k^2v_{ti}^2}\right)\right]=0 \tag{3.29}$$

For $k_{De}^2/k^2 \gg 1$, (3.29) is solved for $\omega = \omega_{sk} + i\gamma_{sk}$ to obtain

$$\omega_{sk} = k\left[\frac{(T_e+3T_i)}{m_i}\right]^{1/2} = kC_s \tag{3.30a}$$

and

$$\gamma_{sk} = -\left(\frac{\pi}{8}\right)^{1/2}\omega_{sk}\left[\frac{C_s}{v_{te}}+\frac{T_e}{T_i}\frac{C_s}{v_{ti}}\exp\left(-\frac{T_e}{2T_i}-\frac{3}{2}\right)\right] \tag{3.30b}$$

where the ion acoustic speed $C_s = [(T_e + 3T_i)/m_i]^{1/2}$ is defined in Chapter 2.

The dispersion relation (3.30a), which is the same as (2.23) derived via the fluid approach, indicates that the ion acoustic wave is non-dispersive. Normally, $T_e \gg T_i$, and thus the ion acoustic speed is governed by the electron temperature and ion mass. The kinetic approach also reveals the ion Landau damping phenomenon. The two terms on the RHS of (3.30b) are ascribed to electron–wave and ion–wave resonance interaction, respectively.

3.3.3 *Magneto plasma*

In the presence of magnetic field, plasma is anisotropic and the dispersion equation (3.25) is not subjected to simple analytic solutions, except in two special cases: wave propagates (1) parallel and (2) perpendicular to the magnetic field. In the first case, wave propagation is not affected by the magnetic field and the results for the electron plasma mode and ion acoustic mode are the same as those presented in (3.28) and (3.30) derived in the unmagnetized case. In the following, we solve (3.25) for wave propagation perpendicular to the magnetic field, i.e., $\mathbf{k} = \hat{\mathbf{x}}, \mathrm{k}, \mathrm{k}_{\|} = 0$, and $\xi_{\mathrm{an}} \to \infty$. In this case, $\xi_{\mathrm{a0}} Z(\xi_{\mathrm{an}}) = -\omega/(\omega - \mathrm{n}\Omega_{\mathrm{a}})$ and (3.25) becomes

$$1 - \sum_{\mathrm{a=e,i}} \frac{\omega_{\mathrm{pa}}^2}{\mathrm{k}^2 \mathrm{v}_{\mathrm{ta}}^2} \sum_{\mathrm{n}=-\infty}^{\infty} \left[\frac{\mathrm{n}\Omega_{\mathrm{a}} \Lambda_{\mathrm{n}}(\beta_{\mathrm{a}})}{(\omega - \mathrm{n}\Omega_{\mathrm{a}})} \right] = 0 \tag{3.31}$$

where the relation $\sum_{\mathrm{n}=-\infty}^{\infty} \Lambda_{\mathrm{n}}(\beta_{\mathrm{a}}) = 1$ is employed.

3.3.3.1 *Upper hybrid mode*

In the high-frequency domain, ion response is ignored; we use the relation $\Lambda_{-\mathrm{n}}(\beta_{\mathrm{a}}) = \Lambda_{\mathrm{n}}(\beta_{\mathrm{a}})$ to rearrange the summation, then (3.31) becomes

$$1 - \frac{\omega_{\mathrm{pe}}^2 \Omega_{\mathrm{e}}^2}{\mathrm{k}^2 \mathrm{v}_{\mathrm{te}}^2} \sum_{\mathrm{n}=1}^{\infty} \left[\frac{2\mathrm{n}^2 \Lambda_{\mathrm{n}}(\beta_{\mathrm{e}})}{(\omega^2 - \mathrm{n}^2 \Omega_{\mathrm{e}}^2)} \right] = 0 \tag{3.32a}$$

In cold plasma, (3.32a) reduces to $1 - [\omega_{\mathrm{pe}}^2/(\omega^2 - \Omega_{\mathrm{e}}^2)] = 0$; it leads to the upper hybrid resonance at $\omega = \omega_{\mathrm{uH}} = (\omega_{\mathrm{pe}}^2 + \Omega_{\mathrm{e}}^2)^{1/2}$. Include the thermal effect but $\mathrm{kv}_{\mathrm{te}} < |\Omega_{\mathrm{e}}|$ and $\omega \neq \mathrm{n}|\Omega_{\mathrm{e}}|$ are assumed, thus only $\mathrm{n} = 1$ and 2 terms in the summation of (3.32a) are included and $\Lambda_1(\beta_{\mathrm{e}}) \sim (\beta_{\mathrm{e}}/2)(1 - \beta_{\mathrm{e}})$ and $\Lambda_2(\beta_{\mathrm{e}}) \sim \beta_{\mathrm{e}}^2/8$ are applied; (3.32a) becomes

$$1 - \frac{\omega_{\mathrm{pe}}^2}{(\omega^2 - \Omega_{\mathrm{e}}^2)} \left[1 + \frac{3\mathrm{k}^2 \mathrm{v}_{\mathrm{te}}^2}{(\omega^2 - 4\Omega_{\mathrm{e}}^2)} \right] = 0 \tag{3.32b}$$

which is then solved to be

$$\omega_{\mathrm{uk}} = \left[\omega_{\mathrm{uH}}^2 + \frac{3\mathrm{k}^2 \mathrm{v}_{\mathrm{te}}^2 \omega_{\mathrm{pe}}^2}{\omega_{\mathrm{pe}}^2 - 3\Omega_{\mathrm{e}}^2} \right]^{1/2} \quad \text{for } \frac{3\Omega_{\mathrm{e}}^2}{\omega_{\mathrm{pe}}^2} < 1 \tag{3.32c}$$

where $12\mathrm{k}^2 \mathrm{v}_{\mathrm{te}}^2 \omega_{\mathrm{pe}}^2/(\omega_{\mathrm{pe}}^2 - 3\Omega_{\mathrm{e}}^2)^2 \ll 1$ is assumed.

Equation (3.32c) is slightly different from (2.24), derived via the fluid approach, in the adiabatic compression factor γ_{e}, which increases from 3 to $3/(1 - 3\Omega_{\mathrm{e}}^2/\omega_{\mathrm{pe}}^2)$. This is because the magnetic field stops the thermal diffusion across it to enhance pressure variation in adiabatic compression.

3.3.3.2 *Electron Bernstein mode*

In thermal plasma with strong magnetic field, $3\Omega_e^2/\omega_{pe}^2 \gg 1$, collective modes with frequencies close to the harmonics of the electron cyclotron frequency also exist. Let $\omega \approx n|\Omega_e|$ for $n \geq 1$, then (3.32a) is reduced to

$$1 - \frac{\omega_{pe}^2}{|\Omega_e|}\frac{[n\Lambda_n(\beta_e)/\beta_e]}{(\omega - n|\Omega_e|)} \approx 0 \tag{3.33a}$$

It is solved to obtain

$$\omega = n|\Omega_e| + \frac{\omega_{pe}^2}{|\Omega_e|}\left[\frac{n\Lambda_n(\beta_e)}{\beta_e}\right] \tag{3.33b}$$

It is noted that in cold plasma, (3.32a) only has a solution $\omega = \omega_{uH}$; thus, (3.33b) does not hold in cold plasma. In other words, electron Bernstein mode exists only in thermal plasma; finite Larmour radius (i.e., $R_L = v_{te}/|\Omega_e| \neq 0$) is essential to the harmonic cyclotron resonance interaction.

3.3.3.3 *Lower hybrid mode*

In the frequency domain that $|\Omega_i| \ll \omega \ll |\Omega_e|$, (3.31) is reduced to

$$1 + \frac{\omega_{pe}^2}{\Omega_e^2}\frac{[1 - \Lambda_0(\beta_e)]}{\beta_e} - \frac{\omega_{pi}^2}{\omega^2}\sum_{n=1}^{\infty}\left[\frac{2n^2\Lambda_n(\beta_i)}{\beta_i}\right] = 0 \tag{3.34a}$$

We now use the relations:

$$\frac{2n^2\Lambda_n(\beta_i)}{\beta_i} = (1/2)\beta_i[\Lambda_{n-2}(\beta_i) + \Lambda_{n+2}(\beta_i) - 2\Lambda_n(\beta_i)] + [\Lambda_{n-1}(\beta_i) + \Lambda_{n+1}(\beta_i)]$$

and

$$\sum_{n=-\infty}^{\infty}\Lambda_n(\beta_a) = 1$$

to obtain

$$\sum_{n=1}^{\infty}2n^2\Lambda_n(\beta_i)/\beta_i = \sum_{n=-\infty}^{\infty}n^2\Lambda_n(\beta_i)/\beta_i$$

$$= 1/2 \sum_{n=-\infty}^{\infty} \{(1/2)\beta_i[\Lambda_{n-2}(\beta_i) + \Lambda_{n+2}(\beta_i) - 2\Lambda_n(\beta_i)] \\ +[\Lambda_{n-1}(\beta_i) + \Lambda_{n+1}(\beta_i)]\} = 1$$

It reduces (3.34) to $1+(\omega_{pe}^2/\Omega_e^2)[1-\Lambda_0(\beta_e)]/\beta_e-(\omega_{pi}^2/\omega^2) = 0$, which is then solved to obtained

$$\omega_{LK} = \omega_{LH}\left(1 + 3/4\frac{k^2 v_{te^2}}{\Omega_e^2}\right)^{1/2} \sim (\omega_{LH}^2 + 3k^2 v_s^2/4)^{1/2} \quad (3.34b)$$

where the lower hybrid resonance frequency $\omega_{LH} = [\omega_{pi}^2\Omega_e^2/(\omega_{pe}^2 + \Omega_e^2)]^{1/2}$ is defined in Chapter 2 and $v_s = (T_e/m_i)^{1/2}$. Again, (3.34b) is slightly different from (2.27) in the case of $\zeta = 0$, in the adiabatic compression factor γ_e and γ_i, which decrease from 1 to 3/4 and from 3 to 0, respectively. In the strong magnetic field case, i.e., $\omega_{pe}^2 \ll \Omega_e^2$, electron plasma becomes uniaxial in the magnetic field direction and only ion plasma is resonant with the transverse disturbance oscillating at the ion plasma frequency, i.e., $\omega_{LH} \sim \omega_{pi}$. On the other hand, in the dense plasma with $\omega_{pe}^2 \gg \Omega_e^2$, Debye shielding keeps the low-frequency disturbance to be quasi-neutral, i.e., electrons and ions move together by the perturbed electric field. Assuming that a perturbed electric field $\mathbf{E} = \hat{\mathbf{x}}E_0e^{-i\omega t}$ is present, where $\Omega_i \ll \omega \ll |\Omega_e|$, the ion and electron quiver velocities are determined to be $V_{ix} = ie E_0/m_i\omega$ and $V_{ex} = -i\omega eE_0/m_e(\omega^2 - \Omega_e^2)$; as shown, $V_{ix} = V_{ex}$ when $\omega = -m_e(\omega^2 - \Omega_e^2)/m_i\omega$, which leads to $\omega \sim (|\Omega_e|\Omega_i)^{1/2} \sim \omega_{LH}$.

3.3.3.4 *Ion Bernstein mode*

In the vicinity of the harmonics of the ion cyclotron frequency, i.e., $\omega \sim n\Omega_i$ with $n \geq 2$, (3.31) becomes

$$1 + \frac{\omega_{pe}^2}{\Omega_e^2} + \frac{\omega_{pi}^2}{\Omega_i^2}\frac{1}{\beta_i}\left\{\left[1 - \Lambda_0(\beta_i)\right.\right. \\ \left.\left. - \left[\frac{2n^2\Lambda_1(\beta_i)}{(n^2-1)} + \frac{n\Omega_i\Lambda_n(\beta_i)}{(\omega - n\Omega_i)}\right]\right]\right\} = 0 \quad (3.35a)$$

The dispersion relation of the ion Bernstein mode is derived to be

$$\omega = n\Omega_i \left\{ 1 + \frac{\Lambda_n(\beta_i)}{\left[1 - \Lambda_0(\beta_i) - \frac{2n^2\Lambda_1(\beta_i)}{(n^2-1)} + \frac{\omega_{uH}^2\Omega_i}{\omega_{pe}^2\Omega_e}\beta_i\right]} \right\}$$

$$\sim n\Omega_i\left[1 - \frac{(1-n^{-2})\Lambda_n(\beta_i)}{\beta_i}\right] \quad \text{for } \beta_i \ll 1$$

$$\sim n\Omega_i\{1 + \Lambda_n(\beta_i)] \quad \text{for } \beta_i \gg 1 \qquad (3.35b)$$

Again, ion Bernstein mode exists only in thermal plasma because harmonic cyclotron resonance interaction relies on finite Larmour radius effect.

3.4 Electromagnetic and Hybrid Wave Dispersion Relations in Cold Magneto Plasma

In cold plasma, we use the extremes $\beta_a \to 0$ and $|\xi_{an}| \to \infty$ to determine

$$\sum_{n=-\infty}^{\infty} [\xi_{a0} Z(\xi_{an}) \overleftrightarrow{\Pi}]$$

in (3.22); it reduces the dielectric tensor to be the same as (2.18a), which is duplicated here

$$\overleftrightarrow{\boldsymbol{\varepsilon}}_0(\mathbf{k},\omega) = \varepsilon_1\hat{\mathbf{x}}\hat{\mathbf{x}} + i\varepsilon_2(\hat{\mathbf{x}}\hat{\mathbf{y}} - \hat{\mathbf{y}}\hat{\mathbf{x}}) + \varepsilon_1\hat{\mathbf{y}}\hat{\mathbf{y}} + \varepsilon_3\hat{\mathbf{z}}\hat{\mathbf{z}} \qquad (3.36)$$

where

$$\varepsilon_1 = 1 - \sum_{a=e,i} \frac{\omega_{pa}^2}{\omega^2 - \Omega_a^2}, \quad \varepsilon_2 = -\sum_{a=e,i} \frac{\Omega_a \omega_{pa}^2}{\omega(\omega^2 - \Omega_a^2)},$$

$$\varepsilon_3 = 1 - \sum_{a=e,i} \left(\frac{\omega_{pa}}{\omega}\right)^2.$$

Thus, the dispersion equation (2.7) is also derived to be the same as (2.19).

3.4.1 *Wave Propagation parallel to the magnetic field*

In this case, $\mathbf{k} = \hat{\mathbf{z}}\mathrm{k}$ and $\overleftrightarrow{\mathbf{I}}_{\mathbf{T}} = \hat{\mathbf{x}}\hat{\mathbf{x}} + \hat{\mathbf{y}}\hat{\mathbf{y}}$, the three scalar equations of (2.6) are

$$\left(\varepsilon_1 - \frac{\mathrm{k}^2\mathrm{c}^2}{\omega^2}\right)\mathrm{E_x} + \mathrm{i}\varepsilon_2\mathrm{E_y} = 0 \tag{3.37a}$$

$$-\mathrm{i}\varepsilon_2\mathrm{E_x} + \left(\varepsilon_1 - \frac{\mathrm{k}^2\mathrm{c}^2}{\omega^2}\right)\mathrm{E_y} = 0 \tag{3.37b}$$

and

$$\varepsilon_3\mathrm{E_z} = 0 \tag{3.37c}$$

and the dispersion equation (2.19) becomes

$$\varepsilon_3\left[\varepsilon_1 + \varepsilon_2 - \frac{\mathrm{k}^2\mathrm{c}^2}{\omega^2}\right]\left[\varepsilon_1 - \varepsilon_2 - \frac{\mathrm{k}^2\mathrm{c}^2}{\omega^2}\right] = 0 \tag{3.38}$$

Equation (3.38) sets up three equations:

3.4.1.1 $[\varepsilon_1 - \varepsilon_2 - (\mathrm{kc}/\omega)^2] = 0$ (*RH circular polarization*)

Equation (3.37c) reduces to $\mathrm{E_z} = 0$, but the set of (3.37a) and (3.37b) has non-trivial solution with $\mathrm{E_y}/\mathrm{E_x} = \mathrm{i}$. Thus, the modes in the dispersion branches presented in the following are RH circularly polarized EM waves.

(a) $\omega > |\Omega_e|$

Ion response is neglected. The $\mathrm{k} - \omega$ dispersion relation given in (2.30) is re-expressed in ω–k dependency as

$$\begin{aligned}\omega &= \frac{|\Omega_e|}{2} + \left[\omega_{pe}^2 + \left(\frac{\Omega_e}{2}\right)^2 + \left(1 - \frac{|\Omega_e|}{\omega}\right)\mathrm{k}^2\mathrm{c}^2\right]^{1/2} \\ &\sim \frac{|\Omega_e|}{2} + \left[\omega_{pe}^2 + \left(\frac{\Omega_e}{2}\right)^2 + \mathrm{k}^2\mathrm{c}^2\right]^{1/2} \quad \text{for } \mathrm{kc} > \omega_{pe} + \frac{|\Omega_e|}{2}\end{aligned} \tag{3.39}$$

This RH circularly polarized EM mode has a cutoff frequency $\omega_{cr} = |\Omega_e|/2 + [\omega_{pe}^2 + (\Omega_e/2)^2]^{1/2}$.

(b) $\Omega_i \ll \omega \leq |\Omega_e|$

The dispersion relation becomes

$$\omega = \frac{|\Omega_e|k^2c^2}{(\omega_{pe}^2 + k^2c^2)}$$

$$\sim k^2c^2\Omega_i/\omega_{pi}^2 \quad \text{for } k^2c^2 \ll \omega_{pe}^2 \tag{3.40a}$$

$$\sim |\Omega_e| \quad \text{for } k^2c^2 \gg \omega_{pe}^2 \tag{3.40b}$$

where (3.40a), which is the same as (2.32), is the dispersion relation of whistler mode; its phase velocity $v_p = (\Omega_i\omega/\omega_{pi}^2)^{1/2}c$ increases with the increase of the wave frequency. In propagation, a whistler pulse disperses in descending tone. It is noted that (3.40a) is also called "Helicon" mode of the electron gas in a solid in the low-frequency domain, where ions are immobile and thus $\Omega_i/|\Omega_e| \to 0$.

The relation (3.40b) gives rise to electron cyclotron resonance. In the magnetic field, electron gyrates, in the RH sense, around the magnetic field at a gyration frequency $|\Omega_e|$. Parallel propagating RH circularly polarized EM wave rotates its fields in the same sense as that of the electrons. When the wave frequency approaches $|\Omega_e|$, wave fields and electrons rotate synchronously. The strong interaction slows down wave propagation to attain electron cyclotron resonance.

(c) $\omega < \Omega_i$

A linear dispersion relation for Alfvén wave is obtained to be

$$\omega = kv_A \tag{3.41a}$$

where the Alfvén velocity, defined in Chapter 2, is

$$v_A = \left(\frac{\Omega_i}{\omega_{pi}}\right)c = \frac{B_0}{(n_0 m_i \mu_0)^{1/2}} = \frac{c}{\sqrt{\varepsilon_A}} \tag{3.41b}$$

It is shown that plasma has a very large relative dielectric constant $\varepsilon_A = (\omega_{pi}/\Omega_i)^2$ in the Alfvén frequency domain. For example, in an argon plasma with plasma density $n_0 = 10^{18}$ m^{-3} and background magnetic field $B_0 = 0.1$ T, $\varepsilon_A \sim 7.6 \times 10^5$.

3.4.1.2 $[\varepsilon_1 + \varepsilon_2 - (kc/\omega)^2] = 0$ (*LH circular polarization*)

Again, (3.37c) reduces to $E_z = 0$; non-trivial solution of (3.37a) and (3.37b) sets $E_y/E_x = -i$, which represents a LH circularly polarized EM wave.

(a) $\omega \gg \Omega_i$

Ion response is neglected; the k$-\omega$ dispersion relation is given in (2.31) which is re-expressed approximately in $\omega-$k dependency as

$$\omega = -\frac{|\Omega_e|}{2} + \left[\omega_{pe}^2 + \left(\frac{\Omega_e}{2}\right)^2 + \left(1 + \frac{|\Omega_e|}{\omega}\right) k^2c^2\right]^{1/2}$$

$$\sim -\frac{|\Omega_e|}{2} + \left[\omega_{pe}^2 + \left(\frac{\Omega_e}{2}\right)^2 + k^2c^2\right]^{1/2} \quad \text{for } kc > \sqrt{2}|\Omega_e| \qquad (3.42)$$

This branch of modes has a cutoff frequency at $\omega_{cl} = -|\Omega_e|/2 + [\omega_{pe}^2 + (\Omega_e/2)^2]^{1/2}$.

(b) $\omega \sim \Omega_i$

The wave is an ion cyclotron resonance mode with

$$\omega \sim \frac{\Omega_i}{(1 + \frac{\omega_{pi}^2}{k^2c^2})} \quad \text{for } k^2c^2 \gg \omega_{pi}^2 \qquad (3.43)$$

As $\omega \to \Omega_i, k \to \infty$, wave is in cyclotron resonance with ions which also gyrate around the magnetic field in the LH sense at ion cyclotron frequency Ω_i.

(c) $\omega \ll \Omega_i$

A linear dispersion relation for Alfvén wave is obtained to be

$$\omega = kv_A \qquad (3.44)$$

In this frequency region, RH and LH circularly polarized Alfvén waves have the same dispersion relation, therefore, (3.44) is also applicable for linear polarization.

3.4.1.3 $\varepsilon_3 = 0$

The set of (3.37a) and (3.37b) has trivial solution $(E_x, E_y) = (0, 0)$ and (3.37c) has non-trivial solution $E_z \neq 0$. The dispersion relation is obtained to be

$$\omega = \omega_{pe} \tag{3.45}$$

This is the electron plasma oscillation, which does not propagate in cold plasma.

3.4.2 *Wave propagation transverse to the magnetic field*

In this case, $\mathbf{k} = \hat{\mathbf{x}}k$ and $\overleftrightarrow{\mathbf{I}}_T = \hat{\mathbf{y}}\hat{\mathbf{y}} + \hat{\mathbf{z}}\hat{\mathbf{z}}$, the three scalar equations of (2.6) are

$$\varepsilon_1 E_x + i\varepsilon_2 E_y = 0 \tag{3.46a}$$

$$-i\varepsilon_2 E_x + \left(\varepsilon_1 - \frac{k^2c^2}{\omega^2}\right) E_y = 0 \tag{3.46b}$$

and

$$\left(\varepsilon_3 - \frac{k^2c^2}{\omega^2}\right) E_z = 0 \tag{3.46c}$$

and the dispersion equation (2.19) becomes

$$\left\{\varepsilon_1 \left[\varepsilon_1 - \frac{k^2c^2}{\omega^2}\right] - \varepsilon_2^2\right\} \left[\varepsilon_3 - \frac{k^2c^2}{\omega^2}\right] = 0 \tag{3.47}$$

3.4.2.1 $\varepsilon_3 - (kc/\omega)^2 = 0$

The set of (3.46a) and (3.46b) has trivial solution $(E_x, E_y) = (0, 0)$, and (3.46c) has non-trivial solution $E_z \neq 0$ which represents a linearly polarized EM mode. The dispersion relation is obtained to be

$$\omega = (\omega_{pe}^2 + k^2c^2)^{1/2} \tag{3.48}$$

which is the same as (2.35). This branch of mode is called ordinary mode (O-mode), which propagates as if plasma is unmagnetized. This wave is cut off from propagation in plasma when $\omega \leq \omega_c = \omega_{pe}$.

3.4.2.2 $\varepsilon_1[\varepsilon_1 - (kc/\omega)^2] - \varepsilon_2^2 = 0$

Equation (3.46c) has trivial solution $E_z = 0$, and the set of (3.46a) and (3.46b) has non-trivial solution setting $E_y/E_x = i\varepsilon_1/\varepsilon_2$, which represents elliptically polarized hybrid wave.

(a) In the frequency domain with $\omega \geq \omega_{cx}$ and $\omega_{c\ell} \leq \omega \leq \omega_{uH}$

The ion terms in the equation can be neglected. At $k = 0, \varepsilon_1 = \pm\varepsilon_2$, and $E_y/E_x = \pm i$ representing RH/LH circular polarization; the corresponding cutoff frequencies are $\omega_{cx,\ell} = \pm|\Omega_e|/2 + [\omega_{pe}^2 + (\Omega_e/2)^2]^{1/2}$, where $\omega_{cx} = \omega_{cr}$.

(1) *X*-mode with $\omega \geq \omega_{cx}$

The k$-\omega$ dispersion relation (2.36) is converted to $\omega-$k dispersion relation approximately to be

$$\omega \sim \frac{|\Omega_e|}{2} + \left[\omega_{pe}^2 + \left(\frac{\Omega_e}{2}\right)^2 + G(\omega, k)k^2c^2\right]^{1/2} \tag{3.49a}$$

$$\sim \frac{|\Omega_e|}{2} + \left[\omega_{pe}^2 + \left(\frac{\Omega_e}{2}\right)^2 + 1/2\left(1 - \frac{|\Omega_e|}{\omega}\right)k^2c^2\right]^{1/2}$$

$$\text{for } \frac{k^2c^2}{\omega|\Omega_e|} \ll 1 \tag{3.49b}$$

$$\sim \frac{|\Omega_e|}{2} + \left[\omega_{pe}^2 + \left(\frac{\Omega_e}{2}\right)^2 + k^2c^2\right]^{1/2}$$

$$\text{for } \frac{k^2c^2}{\omega^2} \to 1 \tag{3.49c}$$

where $G(\omega, k) \sim (1-|\Omega_e|/\omega)[1+(1-|\Omega_e|/\omega)k^2c^2/2\omega|\Omega_e|]/[2+k^2c^2(1-|\Omega_e|/\omega)^2/2\omega|\Omega_e|]$. This branch of modes is called extraordinary mode (X-mode).

(2) *Z*-mode with $\omega_{c\ell} \leq \omega \leq \omega_{uH}$

The dispersion relation has following features

(1) cutoff at $\omega = \omega_{c\ell}$, where $E_y/E_x = -i$; wave is LH circularly polarized EM mode;
(2) at $\omega = \omega_{pe}$, $kc = \omega = \omega_{pe}$, $E_y/E_x = -i|\Omega_e|/\omega$;

(3) as kc $\rightarrow \infty$, $\omega \rightarrow \omega_{\rm uH}$, and $E_y \rightarrow 0$; wave becomes ES mode resonant with electron plasma at upper hybrid resonance frequency $\omega_{\rm uH} = (\omega_{\rm pe}^2 + \Omega_e^2)^{1/2}$.

The k$-\omega$ dispersion relation (2.37) of the Z-mode is re-expressed in the $\omega-$k dependency approximately to be

$$\omega \sim -\frac{|\Omega_e|}{2} + \left[\omega_{\rm pe}^2 + \left(\frac{\Omega_e}{2}\right)^2 + 1/2\left(1+\frac{|\Omega_e|}{\omega}\right)k^2c^2\right]^{1/2}$$

$$\text{for } \frac{k^2c^2}{\omega|\Omega_e|} \ll 1 \tag{3.50a}$$

$$\sim -|\Omega_e|/2 + \left[\omega_{\rm pe}^2 + \left(\frac{\Omega_e}{2}\right)^2 + \left(\frac{\omega_{\rm pe}|\Omega_e|}{\omega^2}\right)k^2c^2\right]^{1/2}$$

$$\text{for } \frac{k^2c^2}{\omega^2} \sim 1 \tag{3.50b}$$

$$\sim \omega_{\rm uH}\left[\frac{\left(1+\frac{\Omega_e^4}{\omega_{\rm uH}^2k^2c^2}\right)}{\left(1+\frac{\Omega_e^2}{k^2c^2}\right)}\right]^{1/2}$$

$$\text{for } \frac{k^2c^2}{\omega^2} \gg 1 \tag{3.50c}$$

Increase k from 0 to infinite, Z-mode, starting as an EM mode, changes to hybrid mode and evolves to electrostatic mode as the frequency approaching $\omega_{\rm uH}$.

(b) In the frequency domain with $\omega < (|\Omega_e|\Omega_i)^{1/2}$
Thus,

$$\varepsilon_1 = 1 - \sum_{a=e,i}\frac{\omega_{\rm pa}^2}{\omega^2-\Omega_a^2} \sim \frac{\omega_{\rm pi}^2(\omega^2-|\Omega_e|\Omega_i)}{|\Omega_e|\Omega_i(\omega^2-\Omega_i^2)} \quad \text{and}$$

$$\varepsilon_2 = -\sum_{a=e,i}\frac{\Omega_a\omega_{\rm pa}^2}{\omega(\omega^2-\Omega_a^2)} \sim \frac{-\omega_{\rm pi}^2\omega}{\Omega_i(\omega^2-\Omega_i^2)}$$

it is shown that $E_y/E_x \sim i(|\Omega_e|\Omega_i - \omega^2)/|\Omega_e|\omega$, representing an elliptically polarized hybrid mode and becoming an ES mode (i.e., $E_y = 0$) as the frequency approaching the lower hybrid resonance frequency $\omega_{LH} = (|\Omega_e|\Omega_i)^{1/2}$; the dispersion relation in the frequency domain with $\omega \ll (|\Omega_e|\Omega_i)^{1/2}$ has the linear dependency

$$\omega = kv_A \tag{3.51}$$

This is the same as that of the Alfvén wave. However, this wave is an elliptically polarized hybrid mode when $\omega \gg \Omega_i$; it turns to an RH circularly polarized hybrid mode (i.e., $E_y/E_x \sim i$) when $\omega \sim \Omega_i$. Only in the frequency domain $\omega \ll \Omega_i$, it becomes a linearly polarized EM mode (i.e., $E_x \cong 0$ and $E_y \neq 0$), which is called compressional Alfvén wave because plasma is compressed by the induced spatially dependent $\mathbf{E}_y \times \mathbf{B}_0$ drift.

3.4.3 *Wave propagation oblique to the magnetic field*

In this general case, $\mathbf{k} = (\hat{\mathbf{x}}\sin\vartheta + \hat{\mathbf{z}}\cos\vartheta)k$ and $\overleftrightarrow{\mathbf{I}}_T = \hat{\mathbf{x}}\hat{\mathbf{x}}\cos^2\vartheta + \hat{\mathbf{y}}\hat{\mathbf{y}} + \hat{\mathbf{z}}\hat{\mathbf{z}}\sin^2\vartheta - (\hat{\mathbf{x}}\hat{\mathbf{z}} + \hat{\mathbf{z}}\hat{\mathbf{x}})\sin\vartheta\cos\vartheta$, the three scalar equations of (2.6) are

$$\left[\varepsilon_1 - \left(\frac{kc}{\omega}\right)^2\cos^2\vartheta\right]E_x + i\varepsilon_2 E_y + \left(\frac{kc}{\omega}\right)^2\sin\vartheta\cos\vartheta E_z = 0 \tag{3.52a}$$

$$-i\varepsilon_2 E_x + \left[\varepsilon_1 - \left(\frac{kc}{\omega}\right)^2\right]E_y = 0 \tag{3.52b}$$

and

$$\left(\frac{kc}{\omega}\right)^2\sin\vartheta\cos\vartheta E_x + \left[\varepsilon_3 - \left(\frac{kc}{\omega}\right)^2\sin^2\vartheta\right]E_z = 0 \tag{3.52c}$$

3.4.3.1 *In the intermediate-frequency domain, $\Omega_i^2 \ll \omega^2 \ll \Omega_e^2$*

Thus, $\varepsilon_1 \sim \omega_{pi}^2(\omega^2 - |\Omega_e|\Omega_i)/(\omega^2|\Omega_e|\Omega_i), \varepsilon_2 \sim -\omega_{pi}^2/\omega\Omega_i$, and $\varepsilon_3 \sim -\omega_{pe}^2/\omega^2$, where $\varepsilon_3^2 \gg \varepsilon_2^2 \gg \varepsilon_1^2$. The dispersion equation (2.19) is reduced to

$$\left(\frac{\omega_{pi}^2}{\omega\Omega_i}\right)^2 - \left(\frac{k_\parallel c}{\omega}\right)^4 = -\left(\frac{k_\perp c}{\omega}\right)^2 \left[\frac{\omega_{pi}^2(\omega^2 - |\Omega_e|\Omega_i)}{\omega^2|\Omega_e|\Omega_i} - \left(\frac{k_\parallel c}{\omega}\right)^2 \frac{\omega_{pe}^2}{(\omega_{pe}^2 + k_\perp^2 c^2)}\right] \quad (3.53)$$

It is solved to obtain a dispersion relation

$$\omega(k, \vartheta) = \frac{k^2c^2|\Omega_e|\left\{\cos^2\vartheta - \sin^2\vartheta\left[\frac{k^2c^2\sin^2\vartheta\cos^2\vartheta}{(\omega_{pe}^2 + k^2c^2\sin^2\vartheta)} - \frac{\omega_{pi}^2}{k^2c^2}\right]\right\}^{1/2}}{\omega_{pe}^2\left[1 + \left(\frac{kc}{\omega_{pe}}\right)^2 \sin^2\vartheta\right]^{1/2}} \quad (3.54)$$

As shown, at $\vartheta = 0$, it reduces to the whistler mode branch $\omega(k, \vartheta = 0) = (k^2c^2\Omega_i/\omega_{pi}^2)$, presented in (3.40a); and at $\vartheta = \pi/2$, it reduces to lower hybrid resonance frequency $\omega(k \to \infty, \vartheta = \pi/2) \sim \omega_{LH}$ with $kc \gg \omega_{pe}$; $\mathbf{E} = \mathbf{E_x}$ is an electrostatic mode.

3.4.3.2 *In the low-frequency domain, $\omega \ll \Omega_i$*

Thus, $\varepsilon_1 \sim \omega_{pi}^2/\Omega_i^2 = \varepsilon_A, \varepsilon_2 \sim \omega\omega_{pi}^2/\Omega_i^3$, and $\varepsilon_3 \sim -\omega_{pe}^2/\omega^2$. First, $|\varepsilon_3| \gg (kc/\omega)^2$ and $E_z \sim 0$ can be assumed. Moreover, $\varepsilon_2 \ll \varepsilon_1$; the dispersion equation is then derived, by combining (3.52a) and (3.52b) with the aid of $\varepsilon_2 \ll \varepsilon_1$, to be

$$\left[\varepsilon_1 - \left(\frac{kc}{\omega}\right)^2 \cos^2\vartheta\right]\left[\varepsilon_1 - \left(\frac{kc}{\omega}\right)^2\right] = 0 \quad (3.55)$$

(a) Compression Alfvén wave:

$$\varepsilon_1 - k^2c^2/\omega^2 = 0 \quad \text{with } E_x \sim 0 \quad \text{and} \quad E_y \neq 0$$

This is a linearly polarized EM mode. The dispersion relation is obtained to be

$$\omega = kv_A \quad (3.56)$$

As shown, this wave is isotropic and propagates at a constant speed v_A. The wave field E_y moves electrons and ions closely together at the drift velocities $\mathbf{V}_{Ee} = \hat{\mathbf{x}}E_y/B_0$ and $\mathbf{V}_{Ei} = \hat{\mathbf{x}}E_y/B_0(1 - \omega^2/\Omega_i^2) \sim \mathbf{V}_{Ee}$, and compress the plasma to produce plasma density variations $\delta n_{e,i} = (k\cos\vartheta/\omega)n_0V_{Ee,i}$. The net charge density perturbation associated with the wave is in the second order $\propto \omega^2/\Omega_i^2$. The wave magnetic field $\mathbf{B} = -(\hat{\mathbf{x}}\cos\vartheta - \hat{\mathbf{z}}\sin\vartheta)\sqrt{\varepsilon_1}E_y/c$ has the time average magnetic energy density $W_m = (1/4)\,B^2/\mu_0 = (1/4)\,\varepsilon_0\varepsilon_A|E_y|^2 \gg (1/4)\,\varepsilon_0|E_y|^2$ because ε_A in this frequency region is much larger than one, indicating that $c|B| \gg |E_y|$. Thus, the wave magnetic field is supported by the induced space charge current rather than the displacement current. In other words, the space charge current density en_0v_{iy} is much larger than the displacement current density $-i\omega\varepsilon_0E_y$, their ratio $|en_0v_{iy}/\omega\varepsilon_0E_y| \sim \omega_{pi}^2/\omega\Omega_i \gg 1$. Thus the displacement current term (the second term) on the RHS of (1.22b) can be neglected when the Alfvén wave is considered. The wave electric energy density, $W_e = 1/4\,\varepsilon_0\varepsilon_A|E_y|^2 = W_m$ is mainly stored in plasma in the kinetic energy of the ion drift motion, which effectively slows down the wave propagation.

(b) Shear Alfvén wave:

$$\varepsilon_1 - k^2c^2\cos^2\vartheta/\omega^2 = 0 \quad \text{with } E_y \sim 0 \quad \text{and} \quad E_x \neq 0$$

This is a linearly polarized hybrid mode. The dispersion relation is obtained to be

$$\omega = kv_A\cos\vartheta = k_zv_A \tag{3.57}$$

As shown, this wave is direction-dependent. The energy of the wave only flows along the magnetic field at a constant speed v_A. The wave field E_x induces a shear motion of the plasma in the y-direction with the drift velocities $\mathbf{V}_{Ee} = -\hat{\mathbf{y}}E_x/B_0$ and $\mathbf{V}_{Ei} = -\hat{\mathbf{y}}E_x/B_0$ $(1 - \omega^2/\Omega_i^2) \sim \mathbf{V}_{Ee}$. Again, in the wave electric field (x−) direction, the displacement current is much smaller than the induced space charge current which supports the wave magnetic field; the wave electric field has a space charge field component, which renders the shear Alfvén wave a hybrid mode. Shear Alfvén wave does not appear at $\vartheta = \pi/2$; at this angle, some of the adopted approximations in the analysis do not valid.

Alfvén wave may be considered to be the propagation of the background magnetic field perturbation. When the magnetic lines of force are distorted, the resultant magnetic stress tends to restore the equilibrium which evolves to oscillatory modes similar to those appearing in a violin-string. Plasma is a highly conducting fluid along the magnetic field; it tends to stick to the magnetic lines of force. When a low-frequency transverse electric field $\mathbf{E}$ appears in the magneto plasma, it forces plasma to drift at the drift velocity $\mathbf{V}_E = \mathbf{E} \times \mathbf{B}_0/B_0^2$ across the magnetic lines of force; the mass motion distorts the magnetic field $\mathbf{B}_0$. The perturbed magnetic field $\mathbf{B}$ produces a magnetic stress $B^2/2\mu_0$ to balance the kinetic pressure of the plasma $(1/2)\, n_0 m_i V_E^2$, i.e., $B^2/2\mu_0 = (1/2)\, n_0 m_i V_E^2$. This balance leads to $B = \sqrt{\varepsilon_A} E/c$, which becomes $E/B = c/\sqrt{\varepsilon_A} = v_A$, showing that this oscillatory mode propagates at the Alfvén velocity v_A.

3.5 Wave Energy Density in Plasma

A wave propagates in plasma, its electric field, $\mathbf{E}(\mathbf{r}, t) = \mathbf{E}_1(t)\, e^{i(\mathbf{k}\cdot\mathbf{r} - \omega t)} + \text{c.c.}$, perturbs the velocity distribution of the plasma, which becomes $f_a(\mathbf{r}, \mathbf{v}, t) = f_{a0}(\mathbf{v}) + f_{a1}(\mathbf{r}, \mathbf{v}, t)$, and to add momentum and kinetic energy to the plasma, where $f_{a1}(\mathbf{v}, \mathbf{r}, t)$ is governed by (3.1). The added momentum density $\mathbf{P}$ and kinetic energy K are governed by the equations

$$\begin{aligned}\frac{d\mathbf{P}}{dt} &= \text{Re}\left\{q_a \left\langle \mathbf{E} \int f_a(\mathbf{r}, \mathbf{v}, t) d\mathbf{v} \right\rangle\right\} \\ &= \text{Re}\left\{q_a \left\langle \mathbf{E} \int f_{a1}(\mathbf{r}, \mathbf{v}, t) d\mathbf{v} \right\rangle\right\} \end{aligned} \tag{3.58}$$

$$\begin{aligned}\frac{dK}{dt} &= \text{Re}\left\{q_a \left\langle \mathbf{E} \cdot \int \mathbf{v} f_a(\mathbf{r}, \mathbf{v}, t) d\mathbf{v} \right\rangle\right\} \\ &= \text{Re}\left\{q_a \left\langle \mathbf{E} \cdot \int \mathbf{v} f_{a1}(\mathbf{r}, \mathbf{v}, t) d\mathbf{v} \right\rangle\right\} \end{aligned} \tag{3.59}$$

where $\langle..\rangle$ stands for time average over the wave period, $\langle\mathbf{E}\rangle = 0$,

$$\begin{aligned} \mathrm{f_{a1}}(\mathbf{r}, \mathbf{v}, \mathrm{t}) &= \mathrm{F_{a1}}(\mathbf{v})\mathrm{e}^{\mathrm{i}(\mathbf{k}\cdot\mathbf{r}-\omega \mathrm{t})} + \mathrm{c.c.} \\ &= -\frac{\mathrm{q_a}}{\mathrm{m_a}}\int_{-\infty}^{\mathrm{t}} \boldsymbol{\nabla}_{\mathrm{v}'}\mathrm{f_{a0}}(\mathbf{v}')\cdot\mathbf{E}_1(\mathrm{t}')\mathrm{e}^{\mathrm{i}(\mathbf{k}\cdot\mathbf{r}'-\omega\mathrm{t}')}\mathrm{dt}' + \mathrm{c.c.} \end{aligned} \tag{3.60}$$

and $\mathbf{E}_1$ is a slow time varying function.

We now expand $\mathbf{E}_1(\mathrm{t}') \cong \mathbf{E}_1(\mathrm{t}) - (\mathrm{t}-\mathrm{t}')\mathrm{d}\mathbf{E}_1(\mathrm{t})/\mathrm{dt}$ in (3.60), where the higher order expansion terms are neglected, to obtain

$$\begin{aligned} \mathrm{F_{a1}}(\mathbf{v}) = -\frac{\mathrm{q_a}}{\mathrm{m_a}}&\left[\mathbf{E}_1(\mathrm{t}) + \mathrm{i}\frac{\partial}{\partial\omega}\frac{\mathrm{d}\mathbf{E}_1(\mathrm{t})}{\mathrm{dt}}\right] \\ &\cdot\int_{-\infty}^{\mathrm{t}} \boldsymbol{\nabla}_{\mathrm{v}'}\mathrm{f_{a0}}(\mathbf{v}')\mathrm{e}^{\mathrm{i}[\mathbf{k}\cdot(\mathbf{r}-\mathbf{r}')-\omega(\mathrm{t}-\mathrm{t}')]}\mathrm{dt}' \end{aligned} \tag{3.61}$$

Substitute (3.61) into (3.58) and (3.59), it results in

$$\frac{\mathrm{d}\mathbf{P}}{\mathrm{dt}} = 2\omega^{-1}\mathbf{E}_1\mathbf{k}\cdot\mathrm{Re}\left[\overleftrightarrow{\sigma}(\mathbf{k},\omega)\cdot\mathbf{E}_1(\mathrm{t}) + \mathrm{i}\frac{\partial}{\partial\omega}\overleftrightarrow{\sigma}(\mathbf{k},\omega)\cdot\frac{\mathrm{d}\mathbf{E}_1(\mathrm{t})}{\mathrm{dt}}\right] \tag{3.62}$$

$$\frac{\mathrm{dK}}{\mathrm{dt}} = 2\mathbf{E}_1\cdot\mathrm{Re}\left[\overleftrightarrow{\sigma}(\mathbf{k},\omega)\cdot\mathbf{E}_1(\mathrm{t}) + \mathrm{i}\frac{\partial}{\partial\omega}\overleftrightarrow{\sigma}(\mathbf{k},\omega)\cdot\frac{\mathrm{d}\mathbf{E}_1(\mathrm{t})}{\mathrm{dt}}\right] \tag{3.63}$$

where $\overleftrightarrow{\sigma}(\mathbf{k},\omega)$ is a conductivity tensor.

With the aid of the relations

$$\mathrm{Re}[\mathrm{i}\overleftrightarrow{\sigma}(\mathbf{k},\omega)] = \omega\varepsilon_0\overleftrightarrow{\varepsilon}_{\mathbf{r}}(\mathbf{k},\omega)$$

$$\mathrm{Re}[\overleftrightarrow{\sigma}(\mathbf{k},\omega)] = \omega\varepsilon_0\overleftrightarrow{\varepsilon}_{\mathbf{i}}(\mathbf{k},\omega) \quad \text{and}$$

$$\omega\overleftrightarrow{\varepsilon}_{\mathbf{i}}(\mathbf{k},\omega)\cdot\mathrm{E}_1 \cong [\overleftrightarrow{\varepsilon}_{\mathbf{r}}(\mathbf{k},\omega) - \overleftrightarrow{\mathbf{I}}]\cdot\mathrm{d}\mathbf{E}_1/\mathrm{dt}$$

Eqs. (3.62) and (3.63) are integrated to be

$$\mathbf{P} = \varepsilon_0\omega^{-1}\mathbf{E}_1\mathbf{k}\cdot\left\{[\overleftrightarrow{\varepsilon}_{\mathbf{r}}(\mathbf{k},\omega) - \overleftrightarrow{\mathbf{I}}] + \omega\frac{\partial}{\partial\omega}\overleftrightarrow{\varepsilon}_{\mathbf{r}}(\mathbf{k},\omega)\right\}\cdot\mathbf{E}_1(\mathrm{t}) \tag{3.64}$$

$$\mathrm{K} = \varepsilon_0\mathbf{E}_1\cdot\left\{[\overleftrightarrow{\varepsilon}_{\mathbf{r}}(\mathbf{k},\omega) - \overleftrightarrow{\mathbf{I}}] + \omega\frac{\partial}{\partial\omega}\overleftrightarrow{\varepsilon}_{\mathbf{r}}(\mathbf{k},\omega)\right\}\cdot\mathbf{E}_1(\mathrm{t}) \tag{3.65}$$

The electric field of the wave also has an energy density $\varepsilon_0 E_1^2$; moreover, when the wave is an EM mode, it also contains a magnetic field energy density $\mu_0 H_1^2$. Thus the total energy density U of the wave is given to be

$$\begin{aligned} U &= \varepsilon_0 \mathbf{E}_1 \cdot \overleftrightarrow{\mathbf{I}} \cdot \mathbf{E}_1(t) + K + \mu_0 H_1^2 \\ &= \varepsilon_0 \mathbf{E}_1 \cdot \left[\frac{\partial}{\partial\omega}\omega \overleftrightarrow{\varepsilon}_{\mathbf{r}}(\mathbf{k},\omega)\right] \cdot \mathbf{E}_1(t) + \mu_0 H_1^2 \end{aligned} \tag{3.66}$$

Examples:

1. Electrostatic mode:

In this case, $\mathbf{E}_1 = E_1\mathbf{k}/k$ and $\mathbf{k}\cdot\overleftrightarrow{\boldsymbol{\varepsilon}}_r(\mathbf{k},\omega)\cdot\mathbf{k} = k^2\varepsilon_L(\omega) = 0$, and (3.64) to (3.66) become

$$\begin{aligned} \mathbf{P} &= \frac{\varepsilon_0}{\omega}\mathbf{k}\left\{\left[\omega\frac{\partial}{\partial\omega}\varepsilon_L(\omega)\right] - 1\right\} E_1^2 \\ K &= \varepsilon_0\left\{\left[\omega\frac{\partial}{\partial\omega}\varepsilon_L(\omega)\right] - 1\right\} E_1^2 \quad \text{and} \\ U &= \varepsilon_0\left[\omega\frac{\partial}{\partial\omega}\varepsilon_L(\omega)\right] E_1^2 \end{aligned}$$

It shows that $P/K = k/\omega$.

The wave energy density of Langmuir wave in unmagnetized plasma is calculated by setting $\varepsilon_L(\omega) = 1 - \omega_{pe}^2/(\omega^2 - 3k^2v_{te}^2) = 0$; it yields $U = 2\varepsilon_0(\omega^2/\omega_{pe}^2)E_1^2$

2. EM mode:

Consider an EM mode in unmagnetized plasma, $\mathbf{E}_1 \perp \mathbf{k}, \varepsilon_1(\omega) = 1 - \omega_{pe}^2/\omega^2 = k^2c^2/\omega^2$; thus

$$\begin{aligned} \mathbf{P} &= 0, \\ K &= \varepsilon_0(\omega_{pe}^2/\omega^2)E_1^2, \quad \text{and} \\ U &= \varepsilon_0(1 + \omega_{pe}^2/\omega^2)E_1^2 + \mu_0 H_1^2 \end{aligned}$$

Problems

P3.1. Show that the (3.5) is the solution of (3.2).
[Hint: Express $\mathbf{r}(t)$ and $\mathbf{v}(t)$, with the aid of (3.5), in terms of the initial values $\mathbf{r}(0) = \mathbf{r}_0$ and $\mathbf{v}(0) = \mathbf{v}_0$]

P3.2. Show the derivation of Eq. (3.22) from Eq. (3.17).

P3.3. Apply (2.9) and (2.10) to obtain the dispersion relation (3.28a) and the damping rate (3.28b) of the electron plasma wave.

P3.4. Show that Eq. (3.25) reduces to Eq. (3.29) in the ion acoustic frequency regime.

P3.5. Again, apply (2.9) and (2.10) to obtain the dispersion relation (3.30a) and the damping rate (3.30b) of the ion acoustic wave.

P3.6. Consider the resonance interaction of an electron with a Langmuir wave $E(z, t) = E_0 \cos(kz - \omega t)$ when the initial electron velocity $v(0) = v_0$ is close to the phase velocity $v_p = \omega/k$ of the wave, i.e., $v_0 \sim v_p$.

In a frame of reference moving at v_0, the coordinate z and velocity v of the electron are transformed to $z' = z - vt$ and $v' = v - v$, where z' and v' are governed by the equations of motion: $dz'/dt = v'$ and $dv'/dt = -(eE/m_e)\cos(kz' + kv_0 t - \omega t)$.

(1) Show that $v'(t) \sim -(eE_0 t/m_e)\cos kz_0$ at resonance $v_0 \sim \omega/k$, where z_0 is the initial coordinate of the electron.
(2) Calculate the spatially averaged energy gain of resonance electrons.

P3.7. Consider a Langmuir wave $E(z, t) = E_0 \sin(kz - \omega t)$; in a frame of reference moving at the phase speed of the wave, i.e., $V_f = \omega/k$, the coordinate z is transformed to $z' = z - V_f t$, and the wave field is transformed to a static field $E(z') = E_0 \sin kz'$, which introduces a static potential distribution $\phi(z') = (E_0/k)\cos kz'$. In the moving frame, electron velocity becomes $u = v - V_f$.

(1) What are the equations of motion of the electron in this moving frame of reference?

(2) Show that if the electron kinetic energy $(1/2)m_eu^2 < 2eE_0/k$ (which is the maximum potential difference), this electron will be trapped in the wave to execute bounce motion in the potential well imposed by the wave.
(3) Show that the bounce frequency at small excursion amplitude is given by $\omega_B = (ekE_0/m_e)^{1/2}$.

P3.8. Consider a magneto plasma generated in a cylindrical chamber, where solenoids are used to generated axial magnetic field. Normally, plasma has a Gaussian shape density distribution in radial direction. Explain how to use a microwave sweep generator, with the aid of the relation (3.48), to determine what the maximum plasma density is.

P3.9. Show that the compressional Alfvén wave compresses the background magnetic flux density $\mathbf{B}_0 = \hat{\mathbf{z}}B_0$. Consider a special case of compressional Alfvén wave propagating in x-direction with an electric field $\mathbf{E} = \hat{\mathbf{y}}E_0\cos(kx - \omega t)$;

(1) Determine the plasma drift velocity.
(2) Determine the wave magnetic field $\mathbf{H} = \mathbf{B}/\mu_0$.
(3) Express the magnetic flux density in terms of the concentration of the field lines with arrow representing the direction, and use the size and direction of an arrow to represent the drift velocity; plot the total magnetic flux density distribution $\mathbf{B}_t = \mathbf{B}_0 + \mathbf{B}$ and drift velocity distribution in the x–z plane.

P3.10. Show that the shear Alfvén wave induces shear motion of plasma following the vibration of the magnetic field line. A shear Alfvén wave propagates in the $x - z$ plane with an electric field $\mathbf{E} = \hat{\mathbf{x}}E_0\cos(k_\perp x + k_z z - \omega t)$.

(1) Determine the wave magnetic field $\mathbf{H}$
(2) Determine the plasma drift velocity
(3) Plot the total magnetic flux density distribution $\mathbf{B}_t = \mathbf{B}_0 + \mathbf{B}$ and drift velocity distribution in the y–z plane.

P3.11. In Fig. 3.1, resonant electrons are trapped in the potential well of the Langmuir wave; these electrons are bounced

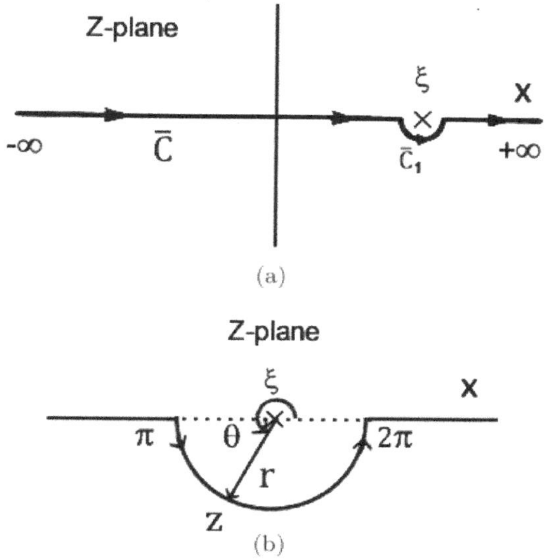

Fig. P3.1. (a) Deformed integration path $\bar{C}$ and (b) enlarged contour $\bar{C}_1$.

back and forth in the potential well. Determine the bounce frequency and its average value.

P3.12. Derive the expansions (3.24a) and (3.24b) of the plasma dispersion function $Z(\xi)$ defined in (3.21), with the aid of the following discussion.

The integrand of the plasma dispersion function (3.21) contains a pole at $x = \xi$. The integration path near the pole has to be deformed to go around the pole; should it go around above the pole or below it. The causality indicates that the deformed path $\bar{C}$ in the complex Z-plane has to go around below the pole as shown in Fig. P3.1a. It is explained as follows:

Starting from an unperturbed state at $t = -\infty$, a perturbation $Ae^{i(\mathbf{k}\cdot\mathbf{r}\omega t)}$ has to grow gradually to reach this finite amplitude A at $t = 0$; in other words, the frequency ω has an infinitesimal positive imaginary, i.e., $\omega = \omega_r + i0^+$; thus the pole of the integrant in (3.21) is inherently above the x-axis.

Therefore, for an analytical function F(x),

$$\int_{-\infty}^{\infty} dx \frac{F(x)}{x-\xi} = \int_{\bar{C}} dz \frac{F(z)}{z-\xi} = P\int_{-\infty}^{\infty} dx \frac{F(x)}{x-\xi} + \int_{\bar{C}_1} dz \frac{F(z)}{z-\xi}$$

where "P" stands for the principal part of the integration, which is expression explicatory to be

$$P\int_{-\infty}^{\infty} dx \frac{F(x)}{x-\xi} = \int_{-\infty}^{\xi-0^+} dx \frac{F(x)}{x-\xi} + \int_{\xi+0^+}^{\infty} dx \frac{F(x)}{x-\xi} \tag{P3.1}$$

and the integration over the contour $\bar{C}_1$ can be carried out analytically. As shown in Fig. P3.1.(b), $z - \xi = re^{i\theta}$, and $dz = i\,re^{i\theta}d\theta$; thus,

$$\int_{\bar{C}_1} dz \frac{F(z)}{z-\xi} = \int_{\pi}^{2\pi} i\,re^{i\theta}d\theta \frac{F(\xi)}{re^{i\theta}} = i\pi F(\xi) \tag{P3.2}$$

The integration of (P3.1) is obtained via expansions of $1/(x-\xi)$ and the aid of the following two integrations:

$$\frac{1}{\sqrt{2\pi}}\int_{-\infty}^{\infty} dx\, x^{2n} \exp\left(-\frac{x^2}{2}\right) = (2n-1)!! \quad \text{and}$$

$$\frac{1}{\sqrt{2\pi}}\int_{-\infty}^{\infty} dx\, x^{-2(n+1)} \exp\left(-\frac{x^2}{2}\right) = \frac{(-1)^n}{(2n+1)!!}$$

$$\text{for } n = 0, 1, 2 \ldots$$

P3.13. Give the ratios of the wave energies of

(1) an O-mode and a X-mode,

(2) an RHCP mode and an LHCP mode.

Chapter 4

Electromagnetic Wave Propagation in the Ionosphere

4.1 Background

The ionosphere is a region of the earth's upper atmosphere where charged particles (electrons and ions) are present. It is located from about 60 km to 1,000 km above the earth's surface. The ionization is mainly caused by the Sun's ultraviolet radiation, and the amount of ionization depends greatly on the amount of radiation received from the Sun. Thus there is a diurnal (time of day) effect and a seasonal effect as well as the sunspot activity which radiates more and has an 11-year cycle. Radiation received also varies with geographical latitude. The ionosphere is non-uniform plasma, and its daytime and nighttime electron density profiles are illustrated in Fig. 4.1.

As indicated in the figure, it is divided into several regions designated by the letters D, E, and F, the latter being subdivided into F1 and F2. The D layer is located lowest among them and has the lowest ionization as well as degree of ionization; it does not have an exact starting height and exists only during the day when this region is in sunlight. E layer is above the D layer and has a peak in its plasma density distribution at about 105 km. In these regions, recombination process in which a free electron is "captured" by a positive ion balances the ionization process to determine the density of ionization present. Therefore, after sunset, the ionization starts to decrease and disappears by night.

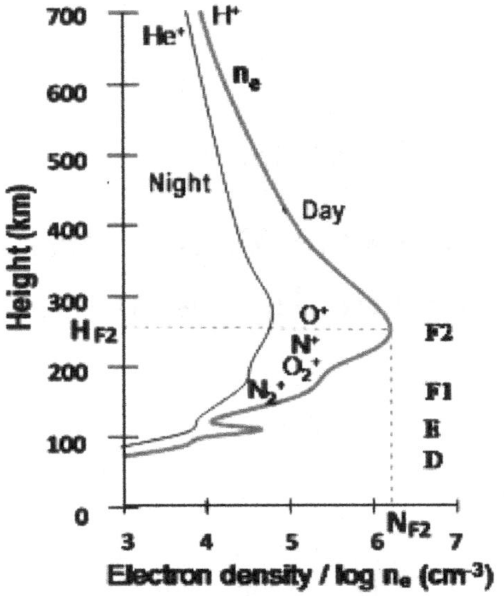

Fig. 4.1. Illustration of daytime and nighttime electron density profiles.

The F layer (or Appleton layer) can be found above the E layer, i.e., ~above 150 km, and it has the highest concentration of charged particles and degree of ionization. It has two parts: the lower F1 layer, and the higher and more electron-dense F2 layer in which there is a density peak located at f_oF2 layer, where f_oF2 represents the maximum plasma frequency f_{pm} of the ionosphere. The f_oF2 layer divides the ionosphere into topside and bottomside. In the D and E regions, plasma is highly (electron–neutral) collisional, the geomagnetic field has insignificant effect on the propagation of high-frequency (HF) waves. On the other hand, the magneto plasma in the F region affects HF and VLF wave propagation strongly. It splits the HF wave into O-mode and X-mode which propagate separately and ducts whistler wave propagation along the magnetic field.

As discussed in Chapter 3, EM wave can be cutoff from propagation when the wave frequency is below the electron plasma frequency. Thus, one can limit the wave frequency to be less than f_{pm} to keep ground transmitted waves in the bottomside of the ionosphere. For instance, ionosphere bounces radio waves down to ground to make radio propagation to distant places on the Earth. Ionosphere may affect Satellite communications. Space weather modifies the ionospheric conditions, which can lead to a total loss of communication due to attenuation and/or severe scintillation when the broadcast signals cross the ionosphere. Scintillation refers to the rapid variation of the amplitude and phase of a received signal, which is caused by undesirable structures appearing in the ionosphere. Severe scintillation conditions can prevent a GPS receiver from locking on to the signal and can make it impossible to calculate a position. Study of wave propagation in the ionosphere has practical importance.

4.2 Wave Propagation in the Bottomside of the Ionosphere

Consider vertical propagation along the density inhomogeneity (z-axis) of the ionospheric plasma, the wave field is governed by the Helmholtz equation

$$\frac{d^2E}{dz^2} + k^2(z)E = 0 \tag{4.1}$$

where $k^2(z) = [\omega_0^2 - \omega_p^2(z)]/c^2$, ω_0 is the wave frequency, $\omega_p(z) = [n_e(z)e^2/m_e\varepsilon_0]^{1/2}$ is the electron plasma frequency, and geomagnetic field is not included in the formulation to simplify the analysis and presentation; the electron density $n_e(z)$ increases with the altitude, thus k(z) decreases as wave propagates upward. Consider the situation that f_0 is less than f_oF2 of the ionosphere (where the plasma density is the maximum). In this case, the wave can reach a layer at $z = z_0$, where $\omega_p(z_0) = \omega_0$ and $k(z_0) = 0$. This point z_0 is called the turning point, where wave reflection occurs. If the wave is not near a turning point z_0 (i.e., reflection layer), WKB solution $E = A_0k^{-1/2}\exp[\pm i\int^z k(z)dz]$ of (4.1) is a good approximation.

4.2.1 *Solution of the wave equation near a turning point*

In the vicinity of $k \to 0$, the electron density can be assumed to have a linearly increasing profile, i.e., $n_e(z) = n_0[1 + (z - z_0)/L]$, where $n_0 = n_e(z_0)$ and L is the linear scale length. Hence, $\omega_p^2(z) = \omega_0^2[1 + (z - z_0)/L]$, and $k^2(z) = [\omega_0^2 - \omega_p^2(z)]/c^2 = \nu(z_0 - z)$, where $\nu = \omega_0^2/Lc^2$. A new coordinate $g = z_0 - z$ is introduced to re-express (4.1) to be

$$\frac{d^2E}{dg^2} + \nu g\, E = 0 \tag{4.2}$$

A coordinate transformation, $y = \frac{2}{3}(\nu g^3)^{1/2}$, leads to $d/dg \to (3\nu y/2)^{1/3} d/dy$ and $d^2/dg^2 \to (3\nu y/2)^{2/3} d^2/dy^2 + (\nu/2)(3\nu y/2)^{-1/3} d/dy$. Introduce another transformation by setting $E = g^{1/2}F$, i.e., $E = \nu^{-1/2}\,(3\nu y/2)^{1/3}F$, (4.2) is transformed to a Bessel equation

$$\frac{d^2F}{dy^2} + \frac{1}{y}\frac{dF}{dy} + \left(1 - \frac{1}{9y^2}\right)F = 0 \tag{4.3}$$

The solution of (4.3) is a Bessel function of order 1/3, i.e., $F = J_{\pm 1/3}(y)$, for $y^2 > 0$, or a modified Bessel function of the second kind of order 1/3, i.e., $F = K_{1/3}(|y|)$, for $y^2 < 0$.

Therefore, the solutions of (4.2) in the two regions $g > 0$ and $g < 0$ are found to be

$$E_\pm = A_\pm g^{1/2} J_{\pm 1/3}\left(\frac{2}{3}\sqrt{\nu} g^{3/2}\right) \quad \text{for } g > 0 \tag{4.4a}$$

and

$$E = D|g|^{1/2} K_{1/3}\left(\frac{2}{3}\sqrt{\nu}|g|^{3/2}\right) \quad \text{for } g < 0 \tag{4.4b}$$

Apply the continuity condition to (4.4) at $g = 0$, i.e., at $z = z_0$, the solution of (4.2) in the region around $z = z_0$ is obtained to be

$$E = \begin{cases} E_0 g^{1/2}\left[J_{1/3}\left(\frac{2}{3}\sqrt{\nu} g^{3/2}\right) + J_{-1/3}\left(\frac{2}{3}\sqrt{\nu} g^{3/2}\right)\right] & \text{for } z \le z_0 \quad (4.5a) \\ E_1 |g|^{1/2} K_{1/3}\left(\frac{2}{3}\sqrt{v}|g|^{3/2}\right) & \text{for } z > z_0 \quad (4.5b) \end{cases}$$

where $E1 = E_0[J_{1/3}(0) + J_{-1/3}(0)]/K_{1/3}(0)$.

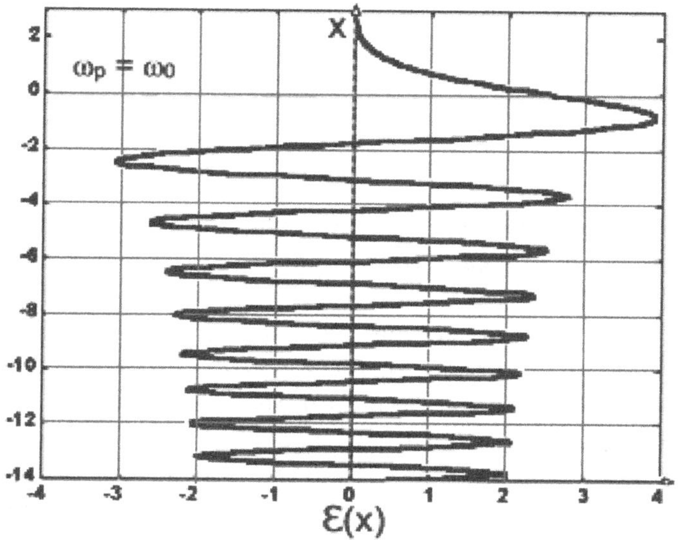

Fig. 4.2. Distribution of the wave electric field near the turning point, where $x = \nu^{1/3}z$.

4.2.2 *Swelling effect of wave propagation to a reflection layer*

Let the turning point $z_0 = 0$ to simplify the expression, so that $g = -z$. Introduce the dimensionless coordinate $x = \nu^{1/3}z = (\omega_0^2/Lc^2)^{1/3}z$, the field function $\varepsilon(x) = E/E_0$ of (4.5) is plotted in Fig. 4.2. As shown, the field amplitude near the turning point (i.e., the reflection height) is enhanced approximately by a factor of 4. This is called the "swelling effect". It results from the total reflection at cutoff, which provides a factor of $\sim$2, and from the slowing down of the wave energy flow while approaching the turning point, which accumulates wave to add another factor of $\sim$2. This enhancement of the HF wave electric field intensity has significant positive effect on exciting plasma instabilities.

4.3 Ray Tracing in Inhomogeneous Media

In inhomogeneous medium, ray tracing is an approach to study wave phenomena. When the wavefront extends uniformly over several wavelengths and the inhomogeneity scale lengths of the medium are large in comparison to the wavelength, wave may be treated as a ray whose trajectory is traced to explore wave propagation.

4.3.1 *A planar stratified medium*

When the medium can be treated to be planar stratified, the Snell's law of refraction is applied to set up the ray trajectory equation. This is illustrated by applying the arrangement shown in Fig. 4.3, in which a medium of continuously varying refractive index $n(z)$ is approximated by a series of plane slabs of thickness Δz, where $\Delta z \to 0$ and each slab i has a uniform refractive index n_i.

As shown in Fig. 4.3, a ray is incident from the medium of refractive index n_0 at an angle θ_0 from the vertical, and the Snell's law $n_j \sin\theta_j = n_{j+1} \sin\theta_{j+1}$ is applied at the interface of two consecutive slabs j and j+1 to determine the moving direction of the ray; $n_0 < n_1 < n_2 \ldots$ is assumed in the plot.

Because the path of the ray in each slab is a straight line, one can identify the relations: $n_i \sin\theta_i = n_j \sin\theta_j$ and $\Delta z/\Delta x_i = \cot\theta_i$, where Δx_i is the horizontal displacement of the ray after transit through slab i. With the aid of these two relations, in the limit of $\Delta z \to 0$, a

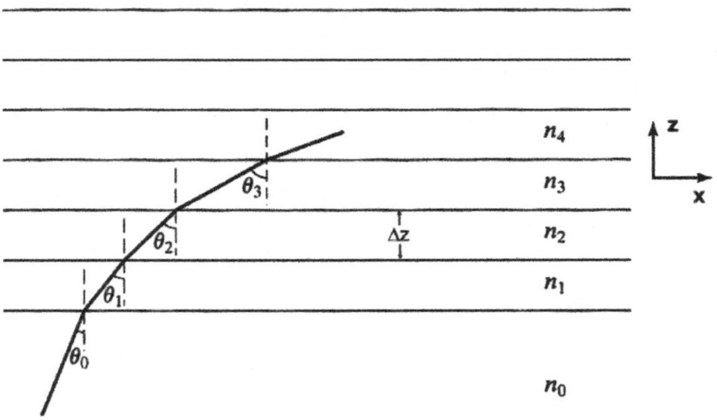

Fig. 4.3. Ray trajectory in a planar stratified medium.

trajectory equation is obtained to be

$$\frac{dz}{dx} = \cot\theta = \frac{(1-\sin^2\theta)^{1/2}}{\sin\theta} = \frac{(n^2 - n_0^2\sin^2\theta_0)^{1/2}}{n_0\sin\theta_0} \tag{4.6}$$

where $n\sin\theta = n_0\sin\theta_0$ is adopted to finalize the equation. If n^2 can be modelled by a second order polynomial, i.e.,$n^2(z) = a + bz + cz^2$, (4.6) can be integrated analytically; otherwise, it can be integrated numerically.

4.3.2 *General formulation*

Consider a general case that wave propagation is governed by a dispersion equation with the generic form

$$G(r, t; k, \omega) = 0 \tag{4.7}$$

We now introduce a generic variable "τ" and take a total τ derivative on (4.7), it yields

$$\frac{dG}{d\tau} = \frac{\partial G}{\partial t}\frac{dt}{d\tau} + \nabla G \cdot \frac{d\mathbf{r}}{d\tau} + \nabla_k G \cdot \frac{d\mathbf{k}}{d\tau} + \frac{\partial G}{\partial \omega}\frac{d\omega}{d\tau} = 0 \tag{4.8}$$

The four terms in (4.8) are arranged into two groups to be

$$\left[\frac{\partial G}{\partial t}\frac{dt}{d\tau} + \frac{\partial G}{\partial \omega}\frac{d\omega}{d\tau}\right] + \left[\nabla G \cdot \frac{d\mathbf{r}}{d\tau} + \nabla_k G \cdot \frac{d\mathbf{k}}{d\tau}\right] = 0 \tag{4.9}$$

where the first group of terms is related to the time variation of the media, while the second group related to the spatial variation of the media. Because (4.9) is obtained from a general approach and the spatial variation and temporal variation are separable, its general solution requires that the following two relations be satisfied simultaneously:

$$\nabla G \cdot \frac{d\mathbf{r}}{d\tau} + \nabla_k G \cdot \frac{d\mathbf{k}}{d\tau} = 0 \tag{4.10}$$

$$\frac{\partial G}{\partial t}\frac{dt}{d\tau} + \frac{\partial G}{\partial \omega}\frac{d\omega}{d\tau} = 0 \tag{4.11}$$

From (4.10) and (4.11), one obtains

$$\frac{d\mathbf{r}}{d\tau} = \nabla_k G \tag{4.12a}$$

$$\frac{d\mathbf{k}}{d\tau} = -\nabla G \tag{4.12b}$$

$$\frac{d\omega}{d\tau} = \frac{\partial G}{\partial t} \tag{4.12c}$$

$$\frac{dt}{d\tau} = -\frac{\partial G}{\partial \omega} \tag{4.12d}$$

In principle, (4.7) can be solved to obtain, for instance, $\omega = \omega(\mathbf{r}, t, \mathbf{k})$; in other words, there are only three independent variables in (4.7). We choose $\mathbf{r}$, t, and $\mathbf{k}$ to be independent variables and ω a dependent variable.

4.3.3 *Ray trajectory equations*

We now take partial derivatives of (4.7) with respective to the three independent variables $\mathbf{k}$, $\mathbf{r}$, and t, respectively, to obtain the following three relationships

$$\nabla_k G + \left(\frac{\partial G}{\partial \omega}\right) \nabla_k \omega = 0 \tag{4.13a}$$

$$\nabla G + \left(\frac{\partial G}{\partial \omega}\right) \nabla \omega = 0 \tag{4.13b}$$

and

$$\frac{\partial G}{\partial t} + \left(\frac{\partial G}{\partial \omega}\right) \frac{\partial \omega}{\partial t} = 0 \tag{4.13c}$$

With the aid of (4.13a) and (4.12d), (4.12a) becomes

$$\frac{d\mathbf{r}}{d\tau} = \nabla_k G = -\left(\frac{\partial G}{\partial \omega}\right) \nabla_k \omega = \frac{d\mathbf{r}}{dt}\left(\frac{dt}{d\tau}\right) = -\left(\frac{\partial G}{\partial \omega}\right) \frac{d\mathbf{r}}{dt},$$

which reduces to

$$\frac{d\mathbf{r}}{dt} = \nabla_k \omega = \mathbf{v}_g \tag{4.14}$$

With the aid of (4.13b) and (4.12d), (4.12b) becomes

$$\frac{d\mathbf{k}}{d\tau} = -\nabla G = -\left(\frac{\partial G}{\partial \omega}\right)\nabla\omega = \frac{d\mathbf{k}}{dt}\left(\frac{dt}{d\tau}\right) = -\left(\frac{\partial G}{\partial \omega}\right)\frac{d\mathbf{k}}{dt},$$

which reduces to

$$\frac{d\mathbf{k}}{dt} = -\nabla\omega \tag{4.15}$$

With the aid of (4.13c) and (4.12d), (4.12c) becomes

$$\frac{d\omega}{d\tau} = \frac{\partial G}{\partial t} = -\left(\frac{\partial G}{\partial \omega}\right)\frac{\partial \omega}{\partial t} = \frac{\partial \omega}{dt}\left(\frac{dt}{d\tau}\right) = -\left(\frac{\partial G}{\partial \omega}\right)\frac{d\omega}{dt},$$

which reduces to

$$\frac{d\omega}{dt} = \frac{\partial \omega}{\partial t} \tag{4.16}$$

Equations (4.14) and (4.15) govern the ray trajectory in the phase space (i.e., $\mathbf{r} - \mathbf{k}$ space). This set of equations (4.14)–(4.16) is essentially the Hamilton's equations of motion with ω and $\mathbf{k}$ to be the Hamiltonian and momentum of the ray.

4.3.4 *Examples*

We now illustrate this ray tracing method with two examples.

4.3.4.1 *Ray trajectory in the bottomside of the ionosphere*

Neglect the geomagnetic field effect, the dispersion relation is given by $\omega = (\omega_p^2 + k^2c^2)^{1/2} = \omega(z, \mathbf{k})$, where $\omega_p^2 = \omega_{p0}^2 z/z_0$, z_0 is the reflection height of the ray, and $\mathbf{k} = \hat{\mathbf{z}}\, k$ with $\hat{\mathbf{z}}$ in the upward direction as shown in Fig. 4.4. Because $\partial\omega/\partial t = 0$, (4.16) leads to $d\omega/dt = 0$; thus, ω is a constant of motion and $\omega_{p0} = \omega$.

With the aid of the explicit function of the dispersion relation, we have $\mathbf{v}_g = \mathbf{k}c^2/\omega$ and $\nabla_z\omega = \hat{\mathbf{z}}\omega_{p0}^2/2z_0\omega$, and (4.14) and (4.15)

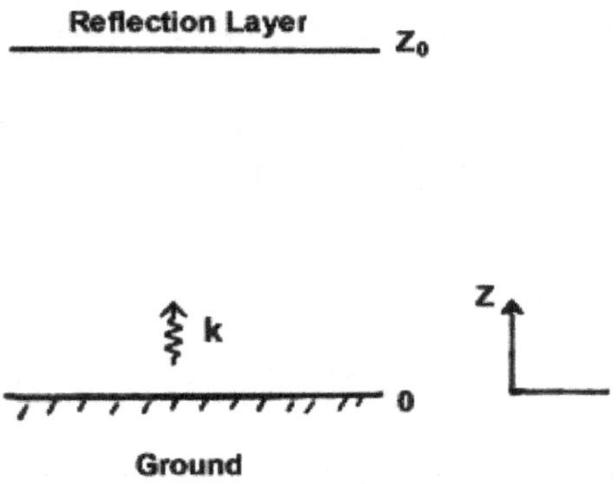

Fig. 4.4. Propagation of a ray from the ground to the reflection layer at $z = z_0$.

become

$$\frac{dz}{dt} = \frac{kc^2}{\omega} \tag{4.17}$$

and

$$\frac{dk}{dt} = -\frac{\omega_{p0}^2}{2z_0\omega} \tag{4.18}$$

These two equations can be combined to a single second-order differential equation

$$\frac{d^2z}{dt^2} = -\frac{c^2\omega_{p0}^2}{2z_0\omega^2} = -\frac{c^2}{2z_0} \tag{4.19}$$

The ray trajectory is determined by solving (4.19) and (4.18), subjected to the initial conditions $z(0) = 0$ and $dz/dt|_{t=0} = c$, and $k(0) = k_0$ (i.e., $\omega/k_0 = c$). In upward propagation period, the trajectory is given by

$$z = z_0\left[1 - \left(1 - \frac{ct}{2z_0}\right)^2\right] \text{ and } k = k_0\left(1 - \frac{ct}{2z_0}\right) \text{ for } t \leq t_0 \tag{4.20}$$

where $t_0 = 2z_0/c$ is the time when the wavenumber k of the ray decreases to zero and the ray is about to be reflected at $z = z_0$.

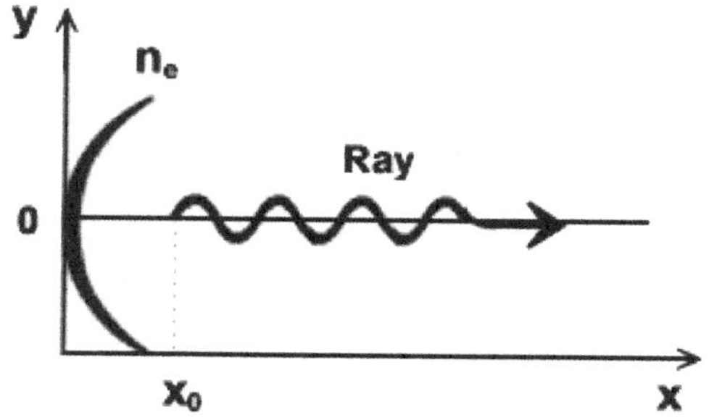

Fig. 4.5. Propagation of a ray along the central axis of a plasma lens that has a parabolic density distribution in the transverse (y-) direction.

For $t > t_0$, the ray starts to propagate downward. The trajectory is given by

$$z = z_0 \left\{ 1 - \left[\frac{c(t - t_0)}{2z_0} \right]^2 \right\} \text{ and}$$

$$k = -k_0 \left[\frac{c(t - t_0)}{2z_0} \right] \text{ for } t_0 \leq t \leq 2t_0 \tag{4.21}$$

4.3.4.2 *Plasma lens*

Consider a plasma lens having a parabolic density distribution, $n_e = n_0(y/y_0)^2$, in y, it guides a ray to propagate along its central (symmetrical) axis in the x-direction as shown in Fig. 4.5.

The dispersion relation is given to be $\omega = [\omega_p^2(y) + k_x^2c^2 + k_y^2c^2]^{1/2} = \omega(y, k_x, k_y)$, where $\omega_p^2 = \omega_{p0}^2 y^2/y_0^2$ and $\omega_{p0} = (n_0e^2/m_e \varepsilon_0)^{1/2}$. Again, (4.16) indicates that ω is a constant of motion. The ray equations (4.14) and (4.15) become

$$\frac{dx}{dt} = \frac{k_x c^2}{\omega} \tag{4.22a}$$

$$\frac{dy}{dt} = \frac{k_y c^2}{\omega} \tag{4.22b}$$

$$\frac{dk_x}{dt} = 0 \tag{4.22c}$$

and

$$\frac{dk_y}{dt} = -\frac{\omega_{p0}^2 y}{y_0^2 \omega} \tag{4.22d}$$

which is subject to the initial conditions $x(0) = x_0, y(0)=0$, and $\mathbf{k}(0) = \hat{\mathbf{x}}k_{x0} + \hat{\mathbf{y}}k_{y0}$, where $k_0 = (k_{x0}^2 + k_{y0}^2)^{1/2} = \omega/c$. First, (4.22c) gives k_x = const. = k_{x0}; thus, we have $k_y = k_{y0}[1 - (\omega_{p0}/k_{y0}c)^2(y/y_0)^2]^{1/2} = k_{y0}[1-(y/y_0)^2]^{1/2}$, where $(\omega_{p0}/k_{y0}c) = 1$, and (4.22a) is integrated to obtain $x = x_0 + k_{x0}c^2t/\omega$. We now take the ratio of (4.22b) to (4.22a), it yields the trajectory equation on the x–y plane to be

$$\frac{dy}{dx} = \frac{k_y}{k_x} = \frac{k_{y0}}{k_{x0}}\left[1 - \left(\frac{\omega_{p0}}{k_{y0}c}\right)^2\left(\frac{y}{y_0}\right)^2\right]^{1/2} \tag{4.23}$$

Equation (4.23) is the same as (4.6), in which $n_0 = 1, \theta_0 = \tan^{-1}(k_{x0}/k_{y0})$, and $n^2 = 1 - (\omega_{p0}/k_0c)^2(y/y_0)^2$. (4.23) is integrated to obtain the ray trajectory y(x) on the x–y plane to be

$$y(x) = \frac{k_{y0}}{k_{x0}}\frac{1}{\kappa_x}\sin[\kappa_x(x - x_0)] \tag{4.24}$$

where $\kappa_x = \omega_{p0}/k_{x0}cy_0$. The ray trajectory (4.24) is plotted in Fig. 4.5. As shown, this ray is guided by the lens to propagate along its central axis.

4.4 Mode Conversion

Waves in plasma are characterized by the frequency and wavelength, as well as by the direction of the wave electric field (with respect to the wave propagation direction) which categorizes the plasma waves into transverse (TEM), longitudinal (ES), and hybrid modes, given in detail in Chapter 2.

4.4.1 *Modes flowing in the spectral regions bounded by the dispersion curves*

The dispersion curves $\omega(k_{||}, k_{\perp})$ of the plasma modes in magneto plasma can be presented in detail in three-dimensional spectral space

($k_{||}$, $k_{\perp}$, ω), where the subscripts || and $\perp$ stand for components parallel and perpendicular to the magnetic field. But a conventional way, which makes easy to describe the physical processes of wave propagation in inhomogeneous magneto plasma, is to plot only the dispersion curves for the 0° (Fig. 2.1) and 90° (Fig. 2.2) propagation angles (with respect to the magnetic field); in these special cases, ω are functions of $|\mathbf{k}|(= k)$. Thus, these curves can be plotted together in a single two-dimensional diagram on the $k-\omega$ plane, as that illustrated in Fig. 4.6(a), and a mode with a propagation angle other than 0° and 90° is placed in the spectral region bounded by a pair of 0° and 90° branches matched through recognizable common characteristic features. In Fig. 4.6(a), only HF branches (in Figs. 2.1 and 2.2) relevant to the following discussion are plotted for "upward propagating modes" in cold plasma embedded in a "downward" geomagnetic field (e.g., the ionosphere of the northern hemisphere). In this case, the left-hand (LH), rather than the right-hand (RH), circularly polarized wave fields rotate in the same direction as the electron gyration. Hence, in Fig. 4.6(a), $\omega_1 = \omega_{cr} = \omega_{cx}$ is the cutoff frequency of the LH circularly polarized (L) branch and extraordinary (X) branch, and $\omega_2 = \omega_{c\ell}$ is the cutoff frequency of the RH circularly polarized (R) branch and the Z-mode branch. $\omega_u = \omega_{uH}$ is the upper hybrid resonance frequency.

Among the curves in Fig. 4.6(a), L and X branches bound the spectral region placing an HF "X-mode" (LH circularly polarized incident wave) with an arbitrary propagation angle between 0° and 90°. Likewise, the R and O pair together with the electron plasma oscillation branch ($\omega = \omega_{pe}$) bound a spectral region for an "O-mode" (RH circularly polarized incident wave) propagating between 0° and 90°. Figure 4.6(b) shows that an O-mode point could escape this bounded spectral region through the intersection point of the R branch and the electron plasma oscillation branch into another spectral region bounded by the R and Z pair together with the electron plasma oscillation branch. When this occurs, an O-mode is converted to a Z-mode. The condition for this linear mode conversion process to occur will be presented in Sec. 4.4.2. The

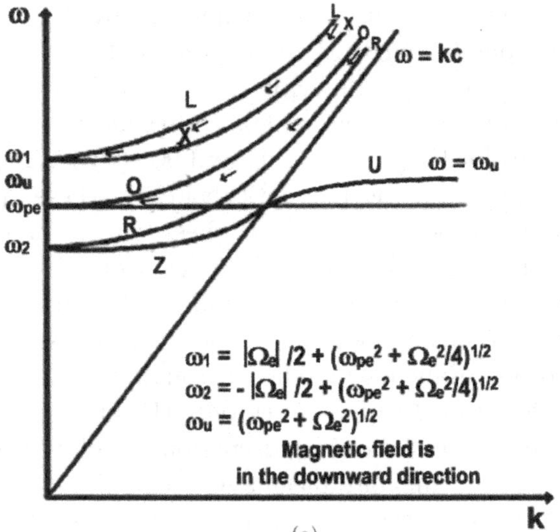

(a)

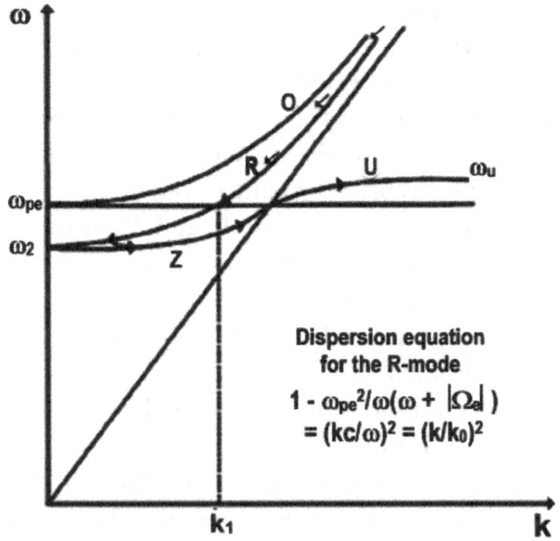

(b)

Fig. 4.6. Dispersion curves of HF EM/hybrid waves propagating parallel (0°) and perpendicular (90°) to the geomagnetic field; (a) arrows indicating the flow of modes, corresponding to the LH/RH circularly polarized, incident waves, in the respective confining spectral region and (b) arrows indicating an RH circularly polarized mode could escape the confining region to change mode type.

polarization of the incident wave at the ground determines its mode type (X or O) and what region in the k–ω plane this initial point is located. LH circularly polarized one is generally named X-mode, and RH circularly polarized one is named the O-mode.

Figure 4.6(a) is plotted for uniform plasma, i.e., in the case of constant electron plasma frequency ω_{pe} and electron cyclotron frequency $|\Omega_e|$. However, the ionosphere is not uniform; although $|\Omega_e|$ may be considered to be a constant, ω_{pe} increases with the altitude in the bottomside of the ionosphere as indicated in Fig. 4.1. When a HF wave at frequency ω_0 propagates upward, the spectral point (k, ω_0) of the wave moves horizontally (i.e., along the k-axis) in a bounded region in Fig. 4.6(a); accordingly, the ω_{pe} in Fig. 4.6(a) moves upward along the vertical (ω) axis, leading all of the dispersion curves to move upward. This will be inconvenient in describing the transition of the mode because Fig. 4.6(a) has to be modified constantly.

However, there is a way to fix ω_{pe} on the vertical axis of Fig. 4.6(a); it is by constantly rescaling the ω axis in the wave propagation. For example, when the wave propagates upward in z, the scale of the ω axis is increasing accordingly to keep $\omega_{pe}(z)$ at the same point on the axis. Consequently, the initial point (k_0, ω_0) of the wave moves downward, but remains in a spectral region bounded by a pair of 0° and 90° branches. The arrows in Fig. 4.6(a) indicate the spectral trajectories $\omega_0(k_{L,R};\ z)$, on the k–ω plane, of vertically incident LH (X-mode) and RH (O-mode) circularly polarized waves. The two incident waves start with the same propagation angle equal to the conjugate α of the magnetic dip angle θ_d, where θ_d is defined to be the angle between the horizontal spatial axis and the geomagnetic field that is inclined downward in the northern hemisphere. An X-mode point (k_L, ω_0), representing a LH circularly polarized incident wave, moves downward with decreasing k_L so that its trajectory stays between the L(0°) and X(90°) curves and meets two curves at the cutoff point (0, ω_1), where $\omega_{pe} = [\omega_0(\omega_0 - |\Omega_e|)]^{1/2} = \omega_{pL}$. On the other hand, the trajectory of the O-mode point, representing an RH circularly polarized incident wave, gradually merges to the O(90°) curve at the cutoff point (0,

ω_{pe}), where $\omega_{pe} = \omega_0 > \omega_{pL}$, located above the reflection height of the LH circularly polarized incident wave.

Next, we show that O-mode wave could convert to ES modes through linear or nonlinear mode conversion in inhomogeneous magneto plasma such as the ionosphere.

4.4.2 *Linear mode conversion*

Linear mode conversion may occur at the intersecting point of two branches of modes. As shown in Fig. 4.6(a) only the O-mode branch intersects with the branch of the electron plasma mode at (k_1, ω_{pe}). Thus the wave has to be transmitted at RH circular polarization and Fig. 4.6(b) will be adopted in the discussion of linear mode conversion. In the figure, only the dispersion curves of the R-, O-, and Z- modes are plotted and the wavevector $\mathbf{k}_1 = \mathbf{k}_{||}$ is in the direction anti-parallel to the geomagnetic field.

The arrows in Fig. 4.6(b) signify the transmission of an RH circularly polarized HF wave, with an oblique incident angle ϑ_0 (with respective to the vertical), into the ionosphere; the vertical component of the wavevector, with the initial value $k_0 \cos\vartheta_0$ where $k_0 = \omega_0/c$ and ω_0 is the wave frequency, decreases continuously as the upward propagating wave encounters the increase of the plasma density. Consequently, the inclined angle ϑ of the propagation, with respective to the vertical, increases continuously. When the wave approaches the O-mode reflection height indicated by the horizontal line $\omega = \omega_{pe}$, the propagation turns toward either (1) the O-curve or (2) the R-curve, depending on the initial inclined angle (i.e., the incident angle). In either cases, the wave will reach an intersecting point at (0, ω_{pe}) or at (k_1, ω_{pe}) on the $\omega = \omega_{pe}$ line. In the first situation, for example, for a vertically incident wave ($\vartheta_0 = 0$), the propagation turns toward the direction perpendicular to the geomagnetic field as the wavevector $\mathbf{k} \to 0$. Because the propagation direction of this wave in the nearby region is drastically different from that of the electron plasma wave (i.e., Langmuir wave), which has a preferred propagation direction parallel to the geomagnetic field, it is not feasible to convert this O-mode wave to a Langmuir wave.

In other words, the intersecting point located at $(k = 0, \omega_{pe} = \omega_0)$ is a turning point (cutoff) rather than a mode conversion point, where the wave is reflected.

The second situation is that the propagation turns to follow the R-curve as shown by the arrows. It requires that the wave be incident at a proper angle. In this case, the wave at the intersecting point $(k_1 = k_{||}, \omega_0 = \omega_{pe})$ propagates anti-parallel to the geomagnetic field and satisfies the R-wave dispersion relation: $1 - \omega_{pe}^2/\omega(\omega + |\Omega_e|) = (kc/\omega)^2$, under the condition $\omega_0 = \omega_{pe}$. This becomes a starting point of a new mode flowing into a spectral region bounded by different dispersion curves. In other words, the wave can propagate continuously upward beyond the $\omega_{pe} = \omega_0$ layer.

Because the ionospheric plasma density is horizontally stratified, the horizontal component, $k_0 \sin\theta_s$ in Fig. 4.7, of the initial wavevector $\mathbf{k}_0$ is conserved in the propagation. Thus, the horizontal component, $k_{||} \sin\alpha$ in Fig. 4.7, of the wavevector $\mathbf{k}_{||}$ at the intersecting point equals to $k_0 \sin\theta_s$, i.e., $k_0 \sin\theta_s = k_{||} \sin\alpha$. Therefore,

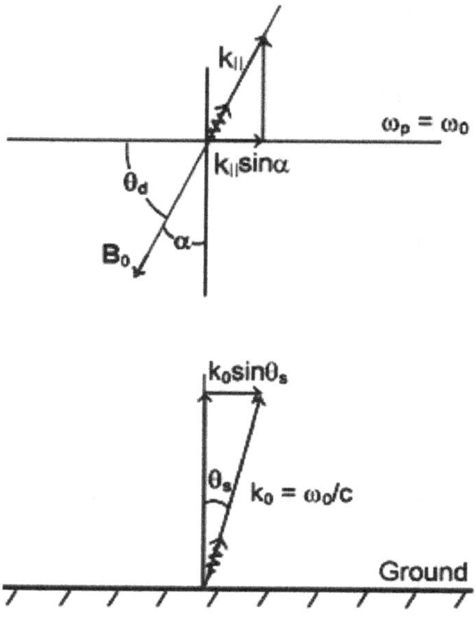

Fig. 4.7. Phase matching condition for mode conversion.

the required incident (Spitze) angle is derived by the phase matching condition $k_0 \sin\theta_s = k_{||} \sin\alpha$, subject to that $(k_{||}, \omega_0 = \omega_{pe})$ satisfy the R-wave dispersion relation. The dispersion relation gives $k_{||} = k_0[|\Omega_e|/(\omega_0 + |\Omega_e|)]^{1/2}$, and the Spitze angle is found to be

$$\theta_s = \sin^{-1}\left\{\left[\frac{|\Omega_e|}{(\omega_0 + |\Omega_e|)}\right]^{1/2} \sin\alpha\right\} \tag{4.25}$$

After passing the $\omega_{pe} = \omega_0$ layer, wave mode enters a different spectral region bounded by the dispersion curves of the R, Z, and electron plasma branches. This new mode continues to propagate upward, but its propagation angle oblique to the downward geomagnetic field decreases continuously from 180°. As the oblique propagation angle decreases to $\pi - \theta_d$ (the vertical component of the wavevector decreases to zero), the wave is reflected at a cutoff layer with $\omega_{pe} \cong \omega_{pR}(1 - k_0^2 \sin^2\theta_s \cos^2\alpha/\omega_0^2)^{1/2}$, where $\omega_{pR} = [\omega_0(\omega_0+|\Omega_e|)]^{1/2}$, and converts to a Z-mode in the region above the O-mode reflection height. This Z-mode wave then propagates downward to further decrease the oblique propagation angle from 180°; as the wave propagates down to the upper hybrid resonance layer at $\omega_{pe} = (\omega_0^2 - \Omega_e^2)^{1/2} = \omega_{pU}$, the oblique propagation angle decreases to 90°, i.e., perpendicular to the magnetic field, and this wave converts to upper hybrid wave.

Through this linear mode conversion, the O-mode HF wave energy will be trapped in the region between $\omega_p = \omega_{pR}$ and ω_{pU} layers.

4.4.3 *Nonlinear mode conversion*

Another effective way (and probably the most effective way) to convert large amplitude EM wave to the ES plasma modes is through parametric instabilities, which excite HF electron waves and low-frequency ion waves simultaneously by the EM wave. This is a nonlinear mode conversion-process employing the nonlinearity of plasma to implement mode–mode couplings and to channel feedbacks to the interactions. The couplings are imposed by the frequency and wavevector matching conditions. The process prefers the excited waves (in particular, the HF electron waves) to be plasma modes to reduce the threshold condition for the commencement of the instability. Thus, the EM wave has to be accessible to the regions where the frequencies of the HF plasma modes are close to the EM frequency. As shown in Fig. 4.6(a), the vertically incident LH circularly polarized wave (X-mode) is reflected at a height with $\omega_{pX} = [\omega_0(\omega_0 - |\Omega_e|)]^{1/2} = \omega_{pL}$ that is below the electron plasma resonance layer $\omega_{pO} = \omega_0$ as well as the upper hybrid resonance layer $\omega_{pU} = (\omega_0^2 - \Omega_e^2)^{1/2}$. On the other hand, the vertically incident RH circularly polarized wave is accessible to both the electron plasma resonance layer and upper hybrid resonance layer, where parametric coupling conditions can be matched. Moreover, near the reflection height the electric field of the O-mode wave is enhanced by a swelling factor of about 4. This "swelling effect" makes it easy to excite parametric instabilities for nonlinear mode conversion and enhances instability growth rates and intensities.

4.5 Mode Coupling

When the plasma waves are generated and grow to large amplitudes, they couple to each other through the nonlinearity of the plasma to generate new plasma waves. In the following, mode coupling of electrostatic modes in unmagnetized plasma is formulated via a kinetic approach. It also illustrates the approach of applying the Eulerian specification of the variables ($\mathbf{r}$, $\mathbf{v}$, t), which are considered to be independent of one another, to analyze the kinetic equation (1.17).

Expend (1.17) with respect to the first-order perturbation force $\mathbf{F}_{a1} = q_a\mathbf{E}_1$ by setting $f_a = f_{a0} + f_{a1} + f_{a2}$ and $\mathbf{F}_a = q_a(\mathbf{E}_1 + \mathbf{E}_2)$, where f_{a0}, f_{a1}, and f_{a2} are the unperturbed background distribution, the first-order (linear) response of the distribution to $\mathbf{F}_{a1}$, and the second-order (nonlinear) response of the distribution to $\mathbf{F}_{a1}$; $\mathbf{E}_2$ is the induced self-consistent (second-order) field responding to f_{a2}. Substitute $f_a = f_{a0} + f_{a1} + f_{a2}$ and $\mathbf{F}_a = q_a(\mathbf{E}_1 + \mathbf{E}_2)$ into (1.17), the zeroth-, first-, and second-order equations are obtained respectively to be

$$\frac{\partial f_{a0}}{\partial t} + \mathbf{v} \cdot \nabla f_{a0} = -\frac{q_a}{m_a} \mathrm{Re}[\langle \mathbf{E}_1 \cdot \nabla_v f_{a1} \rangle] \tag{4.26}$$

$$\frac{\partial f_{a1}}{\partial t} + \mathbf{v} \cdot \nabla f_{a1} = -\frac{q_a}{m_a} \mathbf{E}_1 \cdot \nabla_v f_{a0} \tag{4.27}$$

and

$$\frac{\partial f_{a2}}{\partial t} + \mathbf{v} \cdot \nabla f_{a2} = -\frac{q_a}{m_a} \mathbf{E}_2 \cdot \nabla_v f_{a0} - \frac{q_a}{m_a} \langle \mathbf{E}_1 \cdot \nabla_v f_{a1} \rangle \tag{4.28}$$

where "Re" stands for real part; $\langle \ldots \rangle$ represents a filter keeping only terms having the same phase function as that of the LHS terms of the equation. Neglect the second order term on the RHS of (4.26), in thermal equilibrium the solution of (4.26) is a Maxwellian, $f_{a0} = n_0(m_a/2\pi T_a)^{3/2} \exp(-m_a v^2/2T_a)$, given in (1.19).

In the presence of an electric field disturbance $\mathbf{E}_1(\mathbf{r}, t) = \mathbf{E}_1 \exp[i(\mathbf{k} \cdot \mathbf{r} - \omega t)] + \text{c.c.}$, the linear response in (4.27) is preset to be $f_{a1}(\mathbf{r}, \mathbf{v}, t) = F_{a1}(\mathbf{v}) \exp[i(\mathbf{k} \cdot \mathbf{r} - \omega t)] + \text{c.c.}$., having a similar space–time variation, where $F_{a1}(\mathbf{v})$ is determined, by substituting the preset solution into (4.27), to be

$$F_{a1}(\mathbf{v}) = i\frac{q_a}{m_a} \frac{(\mathbf{E}_1 \cdot \nabla_v f_{a0})}{(\mathbf{k} \cdot \mathbf{v} - \omega)} \tag{4.29}$$

From Gauss's law, $i\mathbf{k} \cdot \mathbf{E}_1 = \rho_1/\varepsilon_0 = \sum_{a=e,i} q_a \int f_{a1}(\mathbf{v}) d\mathbf{v}$, it reduces to

$$\begin{aligned} ikE_1 &= i \sum_{a=e,i} \left(\frac{\omega_{pa}^2}{n_0}\right) \left(\frac{E_1}{k}\right) \int dv \frac{(\partial v_z f_{ao})}{(v_z - \omega/k)} \\ &= -ik \sum_{a=e,i} \chi_a(k, \omega) E_1 \end{aligned} \tag{4.30}$$

where $\mathbf{E}_1$ is set in the z direction without losing the generality and the electric susceptibility χ_a of the species "a" is given to be

$$\chi_a(k,\omega) = \frac{1}{\sqrt{2\pi}}\left(\frac{\omega_{pa}^2}{k^2 v_{ta}^3}\right)\int_{-\infty}^{\infty}\left[\frac{v_z}{\left(v_z - \frac{\omega}{k}\right)}\right]\exp\left(-\frac{m_a v_z^2}{2T_a}\right)dv_z$$
$$= \left(\frac{\omega_{pa}^2}{k^2 v_{ta}^2}\right)[1 + \xi_{a0}Z(\xi_{a0})] \quad (4.31)$$

with $\xi_{a0} = \omega/kv_{ta}$ and $Z(\xi_{a0})$ is the plasma dispersion function (3.21).

Equation (4.30) can be written as $ik[1 + \sum_{a=e,i}\chi_a(k,\omega)]\mathcal{E}_1(k, \omega) = ik\boldsymbol{\varepsilon}_L(k,\omega)\mathcal{E}_1(k,\omega) = 0$, where the relative dielectric response function $\boldsymbol{\varepsilon}_L(k,\omega) = 1 + \sum_{a=e,i}\chi_a(k,\omega)$ and $\boldsymbol{\varepsilon}_L(k,\omega) = 0$ when $\mathcal{E}_1(k,\omega)$ is a mode field. We now proceed to determine the fields $\mathbf{E}_2^{\pm}(\mathbf{r},t) = \mathcal{E}_2^{\pm}(\mathbf{k}_{\pm}, \omega_{\pm})\exp[i(\mathbf{k}_{\pm}\cdot\mathbf{r} - \omega_{\pm}t)] + c.c.$ induced by the coupling of two mode fields $\mathbf{E}_{11}(\mathbf{r},t) = \mathcal{E}_{11}(\mathbf{k}_1, \omega_1)\exp[i(\mathbf{k}_1\cdot\mathbf{r} - \omega_1 t)] + c.c$ and $\mathbf{E}_{12}(\mathbf{r},t) = \mathcal{E}_{12}(\mathbf{k}_2, \omega_2)\exp[i(\mathbf{k}_2\cdot\mathbf{r} - \omega_2 t)] + c.c.$, where $\mathbf{k}_{\pm} = \mathbf{k}_1 \pm \mathbf{k}_2$ and $\omega_{\pm} = \omega_1 \pm \omega_2$. Thus, (4.28) becomes

$$\frac{\partial f_{a2}^{\pm}}{\partial t} + \mathbf{v}\cdot\nabla f_{a2}^{\pm}$$
$$= -\frac{q_a}{m_a}\mathbf{E}_2^{\pm}\cdot\nabla_v f_{a0} - \frac{q_a}{m_a}\langle\mathbf{E}_{11}\cdot\nabla_v f_{a12} + \mathbf{E}_{12}\cdot\nabla_v f_{a11}\rangle \quad (4.32)$$

Substituting $f_{a2}^{\pm} = F_{a2}^{\pm}(\mathbf{v})\exp[i(\mathbf{k}_{\pm}\cdot\mathbf{r} - \omega_{\pm}t)] + c.c.$ into (4.32) yields

$$F_{a2}^{\pm}(\mathbf{v}) = i\frac{q_a}{m_a}\frac{[\mathcal{E}_2^{\pm}\cdot\nabla_v f_{a0} + \mathcal{E}_{11}\cdot\nabla_{\mathbf{v}}F_{a12}^{\pm} + \mathcal{E}_{12}^{\pm}\cdot\nabla_{\mathbf{v}}F_{a11}]}{(\mathbf{k}_{\pm}\cdot\mathbf{v} - \omega_{\pm})}$$
$$= i\frac{q_a}{m_a}\frac{(\mathcal{E}_2^{\pm}\cdot\nabla_{\mathbf{v}}f_{a0})}{(\mathbf{k}_{\pm}\cdot\mathbf{v} - \omega_{\pm})}$$
$$\mp\left(\frac{q_a}{m_a}\right)^2\frac{1}{(\mathbf{k}_{\pm}\cdot\mathbf{v} - \omega_{\pm})}\mathcal{E}_{11}\cdot\left\{\nabla_{\mathbf{v}}\left[\frac{\nabla_{\mathbf{v}}f_{a0}}{(\mathbf{k}_2\cdot\mathbf{v} - \omega_2)}\right]\right\}\cdot\mathcal{E}_{12}^{\pm}$$
$$-\left(\frac{q_a}{m_a}\right)^2\frac{1}{(\mathbf{k}_{\pm}\cdot\mathbf{v} - \omega_{\pm})}\mathcal{E}_{12}^{\pm}\cdot\left\{\nabla_{\mathbf{v}}\left[\frac{\nabla_{\mathbf{v}}f_{a0}}{(\mathbf{k}_1\cdot\mathbf{v} - \omega_1)}\right]\right\}\cdot\mathcal{E}_{11}$$
$$(4.33)$$

where $F^+_{a12} = F_{a12}$ and $F^-_{a12} = F^*_{a12}$; $\mathcal{E}^+_{12} = \mathcal{E}_{12}$ and $\mathcal{E}^-_{12} = \mathcal{E}^*_{12}$; and $F_{a11,2}(\mathbf{v}) = i(q_a/m_a)(\mathcal{E}_{11,2} \cdot \nabla_v f_{a0})/(\mathbf{k}_{1,2} \cdot \mathbf{v} - \omega_{1,2})$, given by (4.29), are substituted.

To simplify the formulation, we consider the case of $\mathcal{E}_{11}||\mathcal{E}_{12}$, i.e., $\mathbf{k}_1||\mathbf{k}_2||\mathbf{k}_\pm$, and set them in the z direction, the last two terms on the RHS of (4.33) can be combined to give

$$F^\pm_{a2}(\mathbf{v}) = i\frac{q_a}{m_a}\left[\frac{\partial v_z f_{a0}}{k_\pm v_z - \omega_\pm}\right]\mathcal{E}^\pm_2 \mp \left(\frac{q_a}{m_a}\right)^2 \times \left\{\frac{1}{(\mathbf{k}_\pm \cdot \mathbf{v} - \omega_\pm)}\partial v_z\left[\frac{(k_\pm v_z - \omega_\pm)\partial v_z f_{a0}}{(k_1 v_z - \omega_1)(k_2 v_z - \omega_2)}\right]\right\}\mathcal{E}_{11}\mathcal{E}^\pm_{12} \tag{4.34}$$

Thus,

$$ik_\pm \mathcal{E}^\pm_2(k_\pm, \omega_\pm) = \sum_{a=e,i}\frac{q_a}{\varepsilon_0}\int F^\pm_{a2}(v)dv$$

$$= -ik_\pm \sum_{a=e,i}\chi_a(k_\pm, \omega_\pm)\mathcal{E}^\pm_2(k_\pm, \omega_\pm) - \sum_{a=e,i}\frac{q_a}{m_a}\left[\frac{k_\pm}{(k_1\omega_2 - k_2\omega_1)^2}\right] \times [k^3_\pm \chi_a(k_\pm, \omega_\pm) - k^3_1\chi_a(k_1, \omega_1) \mp k^3_2\chi_a(k_2, \omega_2)]\mathcal{E}_{11}\mathcal{E}^\pm_{12} \tag{4.35}$$

The second-order field induced by the mode coupling is obtained to be

$$\mathcal{E}^\pm_2(k_\pm, \omega_\pm) = i\sum_{a=e,i}\frac{q_a}{m_a}\frac{1}{(k_1\omega_2 - k_2\omega_1)^2} \times \left\{\frac{[k^3_\pm \chi_a(k_\pm, \omega_\pm) - k^3_1\chi_a(k_1, \omega_1) \mp k^3_2\chi_a(k_2, \omega_2)]}{\varepsilon_L(k_\pm, \omega_\pm)}\right\}\mathcal{E}_{11}\mathcal{E}^\pm_{12} \tag{4.36}$$

As an example, we evaluate the low-frequency fluctuations $(\mathbf{k}_-, \omega_-)$ induced by large amplitude Langmuir waves. In this situation, $\chi_e(\mathbf{k}_{1,2}, \omega_{1,2}) \cong -1$ and $\chi_i(\mathbf{k}_{1,2}, \omega_{1,2}) \cong 0$; $\chi_e(\mathbf{k}_-, \omega_-) \cong$

k_{De}^2/k_-^2; thus,

$$\mathcal{E}_2^-(k_-,\omega_-) = -i\frac{e}{m_e}\left\{\frac{k_-[k_{De}^2 + (k_1^2 + k_1k_2 + k_2^2)]}{(k_1\omega_2 - k_2\omega_1)^2}\right\} \times \frac{\mathcal{E}_{11}\mathcal{E}_{12}^*}{\varepsilon_L(\mathbf{k}_-,\omega_-)} \qquad (4.37)$$

where

$$\omega_- = \omega_1 - \omega_2 = (\omega_{pe}^2 + 3k_1^2v_{te}^2)^{1/2} - (\omega_{pe}^2 + 3k_2^2v_{te}^2)^{1/2} = \frac{3(k_1^2 - k_2^2)v_{te}^2}{2\omega_{pe}}.$$

In the coupling of two parallel propagating Langmuir waves, $\omega_- \gg \omega_{sk} = k_-C_s$, $(\mathbf{k}_-,\omega_-)$ is not an ion acoustic mode and $\varepsilon_L(\mathbf{k}_-,\omega_-) \cong k_{De}^2/k_-^2 \gg 1$; thus, the induced fluctuation field intensity $|\mathcal{E}_2^-(\mathbf{k}_-,\omega_-)|$ is low. However, in the coupling of two oppositely propagating Langmuir waves $(\mathbf{k}_1,\omega_1)$ and $(\mathbf{k}_2,\omega_2)$, which have slightly different wavenumbers, ion acoustic mode $(\mathbf{k}_s,\omega_s)$ may be generated. It requires that $k_s = k_- = |k_1| + |k_2|$ and $\omega_s = k_sv_s = 1.5(k_1^2 - k_2^2)v_{te}^2/\omega_{pe}$, which lead to $|k_1| - |k_2| = 2C_s\omega_{pe}/3v_{te}^2 = 2/3\times (m_e/m_i)^{1/2}(1 + 3T_i/T_e)^{1/2}k_{De}$ and

$$\mathcal{E}_2^-(\mathbf{k}_s,\omega_s) \approx \frac{e}{m_e}\left(\frac{k_s\omega_s}{2\gamma_s\omega_1^2}\right)\mathcal{E}_{11}\mathcal{E}_{12}^* \qquad (4.38)$$

where $\mathbf{k}_s \cong 2\mathbf{k}_1$ and $\gamma_s = -(\pi/8)^{1/2}\omega_s[(m_e/m_i)^{1/2} + (T_e/T_i)^{3/2} \times \exp(-T_e/2T_i - 3/2)] \sim -(\pi/8)^{1/2}\omega_s(m_e/m_i)^{1/2}$ for $T_i/T_e \ll 1$.

The second example is the coupling of Langmuir waves and ion acoustic waves, which broadens the Langmuir spectrum. In (4.36), set $(\mathbf{k}_1,\omega_1)$ and $(\mathbf{k}_2,\omega_2)$ to be the Langmuir and ion acoustic mode, respectively. Because $\omega_2 \ll \omega_1$, it requires that $k_\pm \sim -k_1$; thus, $|k_2| \sim 2|k_1|$ and (k_+,ω_+) is generated by two anti-parallel propagating modes and (k_-,ω_-) is generated by two parallel propagating modes. In this situation, $\chi_e(\mathbf{k}_1,\omega_1) \cong -1 = \chi_e(k_\pm,\omega_\pm)$ and $\chi_i(\mathbf{k}_1,\omega_1) \cong 0 \cong \chi_i(k_\pm,\omega_\pm)$; $\chi_e(\mathbf{k}_2,\omega_2) \cong k_{De}^2/k_2^2 \cong -\chi_i(\mathbf{k}_2,\omega_2)$; $\varepsilon_L(k_\pm,\omega_\pm) \cong i(\pi/2)^{1/2}(k_{De}^2/k_\pm^2)(\omega_\pm/k_\pm v_{te})\exp(-\omega_\pm^2/2k_\pm^2v_{te}^2)$, and

(4.36) reduces to

$$\mathcal{E}_2^{\pm}(k_{\pm},\omega_{\pm}) \cong \frac{1}{\sqrt{2\pi}}\frac{e}{m_e}\left(\frac{k_1^2 v_{te}}{\omega_1^3}\right)\exp\left(\frac{\omega_1^2}{2k_1^2 v_{te}^2}\right)\mathcal{E}_{11}\mathcal{E}_{12}^{\pm} \quad (4.39)$$

So far, we have studied the coupling of two plasma waves, which generates a new plasma mode or drives a fluctuation. On the other hand, the self-coupling of plasma waves, retained on the RHS term of (4.26), modifies the background distribution.

4.6 Quasi-Linear Diffusion and Effective Temperature

We now consider a spectral distribution of plasma waves in the background plasma, the total wave field is given by $\mathbf{E}_1(\mathbf{r},t) = \sum_k \mathcal{E}_{1k}\exp[i(\mathbf{k}\cdot\mathbf{r}-\omega_k t)] + c.c.$, where $\omega_k = \omega_{kr} + i\gamma_k$, and ω_{kr} and γ_k are the real frequency and damping (or growth) rate of the $(\mathbf{k},\omega_k)$ mode. Substituting (4.29) to the RHS of (4.26) yields a diffusion equation

$$\begin{aligned}\frac{\partial f_{a0}}{\partial t} &+ \mathbf{v}\cdot\nabla f_{a0}\\ &= \left(\frac{q_a}{m_a}\right)^2 \nabla_v \cdot \left\{\sum_k \frac{2\gamma_k \mathcal{E}_{1k}\mathcal{E}_{1k}^*}{[(\mathbf{k}\cdot\mathbf{v}-\omega_{kr})^2+\gamma_k^2]}\right\}\cdot\nabla_v f_{a0}\\ &= \nabla_v \cdot \overleftrightarrow{\mathbf{D}}_a(\mathbf{v})\cdot\nabla_v f_{a0} \end{aligned} \quad (4.40)$$

where $\mathbf{D}_a(\mathbf{v}) = (q_a/m_a)^2\sum_k 2\gamma_k\mathcal{E}_{1k}\mathcal{E}_{1k}^*/[(\mathbf{k}\cdot\mathbf{v}-\omega_{kr})^2+\gamma_k^2]$ is the diffusion tensor in velocity space.

To simply the illustration, we consider one-dimension case in a uniform plasma, which reduces (4.40) to

$$\frac{\partial f_{a0}}{\partial t} = \partial_v D_a(v)\partial_v f_{a0} \quad (4.41)$$

where $D_a(v) = (q_a/m_a)^2\sum_k 2\gamma_k|\mathcal{E}_{1k}|^2/[(kv-\omega_{kr})^2+\gamma_k^2]$ is the diffusion coefficient; as shown $|D_i(v)| \ll |D_e(v)|$. In the following, we focus on electron diffusion. $D_e(v)$ is large in the resonance region, around $v = \omega_{kr}/k$, where the diffusion process is called quasi-linear diffusion. It flattens the electron distribution function in the resonance region to form a plateau.

In the bulk region of the distribution, where $v \ll \omega_{kr}/k$, wave and charged particles interact non-resonantly. In the case of Langmuir waves, most of the plasma is in the non-resonance region, and thus, $D_e(v) \cong (e/m_e)^2 \sum_k 2\gamma_k |\mathcal{E}_{1k}|^2/\omega_{kr}^2 = (e/m_e)^2 (d/dt)[\sum_k |\mathcal{E}_{1k}|^2/\omega_{kr}^2]$ and (4.41) becomes

$$\frac{\partial f_{a0}}{\partial t} = \frac{1}{2m_e}\frac{dT_w}{dt}\partial_v^2 f_{e0} \tag{4.42}$$

where $T_w = 2\varepsilon_0 \sum_k (\omega_{pe}^2/\omega_{kr}^2)|\mathcal{E}_{1k}|^2/n_0$ is the time average wave energy per electron. Let $T_{eff} = T_e + T_w$ and the unperturbed distribution $f_{eo}(v, T_w = 0) = n_0(m/2\pi T_e)^{1/2} \exp(-m_e v^2/2T_e)$, one can show that (4.42) has the solution

$$f_{e0}(v, T_w) = n_0 \left(\frac{m_e}{2\pi T_{eff}}\right)^{1/2} \exp\left(-\frac{m_e v^2}{2T_{eff}}\right) \tag{4.43}$$

As shown, the bulk of the electron plasma has an effective temperature T_{eff}; the temperature increment T_w is proportional to the wave energy density. (Thus, it is realized that the Langmuir waves introduce quiver motion of non-resonant electrons) the time average kinetic energy KE of the electron quiver motion in the Langmuir wave fields is $(1/2)m_e \sum_k (2e|\varepsilon_{1k}|/m_e\omega_{kr})^2 = T_w$. When the spectral energy density of the Langmuir wave is large, it can effectively modify the wave dispersion relations.

Problems

P4.1. Show the assumptions employed to obtain WKB solution $E = A_0 k^{-1/2} \exp[\pm i \int^z k(z)dz]$ of Eq. (4.1).

P4.2. Consider an EM wave propagation in plasma of increasing density, near the turning point, then calculate

(1) Phase velocity v_p
(2) Group velocity v_g
(3) The product $v_p v_g$

(4) Wave magnetic field intensity in term of the electric field intensity

P4.3. Derive explicitly the ray tracing equations for an EM wave propagating in inhomogenous unmagnetized plasma.

P4.4. Extend the example (1) Ray trajectory in the bottomside of the ionosphere to the general case of oblique incidence; thus $\mathbf{k} = \hat{\mathbf{x}}k_x + \hat{\mathbf{z}}k_z$ with $\hat{\mathbf{z}}$ in the upward direction and $\hat{\mathbf{x}}$ in the horizontal direction, and with the initial condition $\mathbf{k}(0) = \hat{\mathbf{x}}k_{x0} + \hat{\mathbf{z}}k_{z0} = \mathbf{k}_0$ where $k_0 = (k_{x0}^2 + k_{z0}^2)^{1/2} = \omega/c$. The dispersion relation $\omega = (\omega_p^2 + k_x^2c^2 + k_z^2c^2)^{1/2} = \omega(z, \mathbf{k})$, where $\omega_p^2 = \omega_{p0}^2 z/z_0, z_0$ is the reflection height of the ray and $\omega_{p0}^2 = \omega^2 - k_{x0}^2c^2$; the ray starts at $x(0) = 0$ and $z(0) = 0$.

(1) Show that the trajectory is given by $z/z_0 = 1-(1-\alpha x/z_0)^2$, where $\alpha = \omega_{p0}/k_{x0}c$

(2) Plot the trajectory z/z_0 vs x/z_0, i.e., on the x–z plane, of one round trip for incident angle of 30°.

P4.5. Show that the ray trajectory in Fig. 4.5 bounces back and forth in y-direction at a bounce frequency $\omega_B = (\omega_{p0}c/\omega y_0)$, i.e., $y(t) = y_0 \sin \omega_B t$.

Chapter 5

Electromagnetic Wave Interaction with the Ionosphere

Ionospheric heating by powerful HF waves transmitted from the ground is a platform for experimental and theoretical investigation of wave–wave and wave–particle interactions in magneto plasma. Considerable advancement toward the understanding of nonlinear plasma processes has been recognized through the observations of various heating-induced phenomena at Arecibo, Puerto Rico, Tromso, Norway, Gakona, Alaska, and elsewhere. The facility is also applied to explore new technological innovations, such as an ionospheric very-low-frequency (VLF) transmitter for underwater communications and for conditioning the magnetosphere.

The HF transmitter of the High Frequency Active Auroral Research Program (HAARP) is the most updated facility, which was built in Gakona, Alaska, for conducting ionospheric heating experiments. It has a rectangular planar array of 180 elements. Each element consists of a low band (2.8 to 7.6 MHz) and a high band (7.6 to 10 MHz) crossed dipole antennas. Each crossed dipole radiates circularly polarized wave up to 20 kW. Overall, the HF transmitter at Gakona, Alaska, radiates circularly polarized waves up to 3.6 MW in the frequency band from 2.8 to 10 MHz. The antenna gain, which increases with the radiating frequency, varies from 15 to 30 dB. The effective radiated power (ERP) up to 95 dBw is available for exploring nonlinear plasma physics through the observations of the

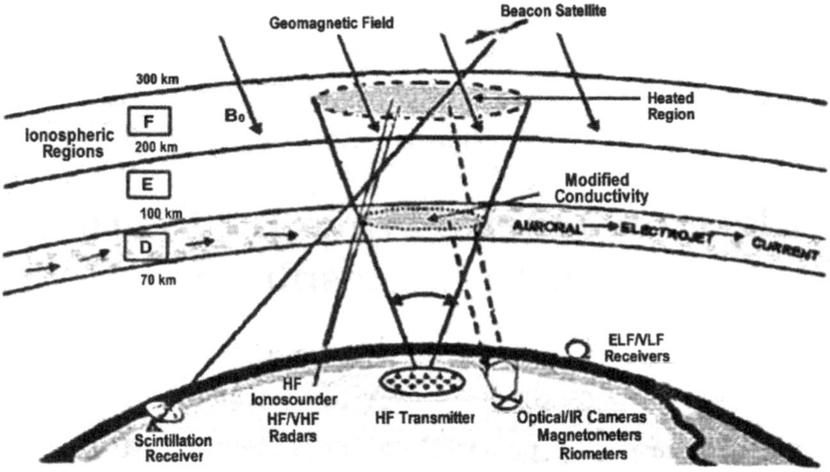

Fig. 5.1. Ionospheric heating experiments and some remote sensing facilities.

heating effects on the bottomside of the ionosphere and for developing ionospheric devices, such as VLF transmitter and infrared generator.

Figure 5.1 is a cartoon showing HAARP heating facility, several diagnostic systems, and anticipated active regions in the ionosphere.

HF heater wave interacts with the charged particles in the ionosphere, and only electrons can effectively respond to the fast oscillation of the HF wave electric field. Neutral particles can also indirectly experience the presence of the HF heater wave through elastic and inelastic collisions with electrons. In the daytime, the bottomside of the ionosphere is low, where the electron–neutral collision frequencies are large; electrons can be heated by the HF heater wave directly. In the nighttime, electrons cannot be heated effectively by the HF heater wave directly in the F region of the ionosphere where the dominant electron–ion collision frequency is low. Due to large mass ratio, only a small fraction of the electron quiver energy in the wave field is transferred to the neutral particles in elastic collisions; thus the temperature elevation of the neutral particles is small. The inelastic collisions which increase the internal energy of the neutral particles involve mainly suprathermal electrons with energy exceeding 2 eV.

Most of the HF heating experiments were focused on the F region modification, where the anomalous heating processes have to be involved in delivering wave energy to plasma electrons and the energy thermal and temperature equilibrium times are much longer. In other words, non-thermal processes were expected to prevail when the F region modifications occurred and new phenomena attributed to wave–electron and wave–wave interactions could be explored.

Because the Coulomb (electron–ion) collisions cannot efficiently damp the HF heater wave transmitted to the F region of the ionosphere, the HF transmitter is usually run at frequency less than the maximum cutoff frequency in the ionosphere to keep the wave energy in the bottomside region of the ionosphere. Moreover, O-mode (rather than X-mode) heater is transmitted so that parametric instabilities can be excited to achieve a fast conversion of the HF heater wave into electrostatic (ES) plasma waves, which are confined in plasma.

The following phenomena were observed when O-mode HF heater waves were transmitted. Heater waves excite parametric instabilities in the F region of the ionosphere to facilitate nonlinear mode conversions and anomalous heating, which evolve plasma to a turbulent state.

5.1 HF-Enhanced Plasma Lines

Backscatter radars receive return signals from incoherent and coherent backscattering, wherein incoherent backscattering is mainly caused by the non-uniformity of the background plasma the and coherent one is attributed to backward Bragg scattering by the plasma waves (electron plasma waves as well as ion plasma waves).

5.1.1 *Bragg scattering*

In the presence of a plasma wave (ω, $\mathbf{k}$), the radar signal (ω_{R}, $\mathbf{k}_{\mathrm{R}}$) is scattered to produce new signals (ω_{Rs}, $\mathbf{k}_{\mathrm{Rs}}$) whose frequencies and wavevectors are imposed by the Bragg scattering (matching) conditions: $\omega_{\mathrm{Rs}} = \omega_{\mathrm{R}} \pm \omega$ and $\mathbf{k}_{\mathrm{Rs}} = \mathbf{k}_{\mathrm{R}} \pm \mathbf{k}$. Because $\omega_{\mathrm{R}} \gg \omega$, $|\mathbf{k}_{\mathrm{Rs}}| \cong$

$|\mathbf{k}_R|$; in the backscattering, $\mathbf{k}_{Rs} \cong -\mathbf{k}_R$ and $\mathbf{k} = \mathbf{k}_p \cong \mp 2\,\mathbf{k}_R$. Thus, the wavelength of the plasma waves that set up coherent backscattering of radar signals has to be half that of the radar signal, i.e., $\lambda = \lambda_R/2$, and to propagate parallel or anti-parallel to the radar signal. The parallel and anti-parallel propagating plasma waves produce frequency down–shifted and up–shifted radar returns, respectively. Offset by the radar frequency, the frequency spectrum of backscatter-radar returns distributes on both sides of the central frequency at zero. The spectral lines on the negative side correspond to up-going plasma waves and those on the positive side correspond to down-going plasma waves. In the unheated quiet ionosphere, the spectral intensity of plasma lines in the backscatter-radar returns is in the noise level.

5.1.2 *HF enhanced plasama lines*

When O-mode HF heater waves $(\omega_0, \mathbf{k}_0 \cong 0)$ were transmitted in heating experiments, both up-going and down-going plasma waves were recorded by backscatter radars. The frequency spectra of the electron plasma lines contain discrete spectral peaks located at HF heater wave frequency ω_0 and at frequencies downshifted from the HF heater wave frequency by $\Delta\omega \cong (2N+1)2k_R C_s, N = 0, 1, 2\ldots$, where $2k_R C_s$ is an ion acoustic frequency and C_s is the ion acoustic speed. These lines were enhanced by the HF heater wave and thus named "HF enhanced plasma lines (HFPLs)". The spectral features suggest that HFPLs are correlated to the oscillating two stream instability (OTSI), parametric decay instability (PDI), and Langmuir cascade. First, these parametric instabilities are only excited by the O mode HF heater wave. OTSI excites Langmuir waves $(\omega, \mathbf{k})$ together with non-oscillatory purely growing modes $(\omega_s \cong 0, \mathbf{k}_s \cong -\mathbf{k})$; thus, the spectral peak of excited Langmuir waves is located at the heater frequency ω_0. PDI decays the HF heater wave to Langmuir waves $(\omega = \omega_0 - \omega_s, \mathbf{k})$ and ion acoustic waves $(\omega_s, \mathbf{k}_s \cong -\mathbf{k})$. The wavevector of the HFPLs is $\mathbf{k}_p \cong \mp 2\mathbf{k}_R$, thus the frequency ω of the spectral peak is downshifted from ω_0 by $\omega_{sp} = k_{sp} C_s = 2k_R C_s$, i.e., $\omega = \omega_0 - 2k_R C_s$. The PDI-excited Langmuir wave $(\omega, \mathbf{k})$ can cascade to a new Langmuir wave ($\omega', \mathbf{k}' \cong -\mathbf{k}$) and an ion acoustic wave ($\omega_s, \mathbf{k_s} \cong 2\mathbf{k}$), and the cascade process may proceed further to generate more cascade lines. The frequency, ω', of the first cascade plasma line is downshifted from ω by about $4k_R C_s$, and thus from ω_0 by about $6k_R C_s = 3\omega_{sp}$.

5.1.3 *Plasma line overshoot*

In Arecibo HF heating experiments, the intensity and the originating height of the HFPLs showed overshoot and downshifting in time, respectively. Thus it, is realized that the relevant parametric instabilities prefer to excite Langmuir waves along the geomagnetic field. On the other hand, the Langmuir waves, giving HFPLs at the Arecibo site, propagate vertically at 40° oblique to the geomagnetic field. These waves are not the preferential ones and are suppressed in time by the dominant ones which propagate at smaller oblique angles with larger growth rates and modify the background. The matching height of the Langmuir wave shifts downward as the effective electron temperature T_{eff} in (4.43) increases; this temperature is proportional to the total spectral intensity of the Langmuir waves.

5.1.4 *Asymmetry of the spectral distribution*

Moreover, the spectral distribution of the HFPLs on the negative side (corresponding to up-going plasma waves) and on the positive side (down-going plasma waves) is asymmetric, and the spectral intensity on the positive side appears to be stronger. Both up-going and down-going plasma waves are excited by the parametric instabilities. Because the plasma density in the bottomside of the ionosphere increases with the altitude, the wavenumber k of the down-going plasma wave is increasing during the downward propagation; on the other hand, the up-going plasma wave decreases its wavenumber k to $k = 0$ at $\omega_p = \omega$, where the wave is reflected to become down-going plasma wave. Because parametric instabilities are excited very close to the reflection height of the plasma waves, the reflected plasma waves do not attenuate much before joining the down-going plasma waves excited directly by the parametric instabilities. These reflected waves, which also backscatter radar signals to produce frequency upshifted radar returns, enhance the spectral intensity of the HFPLs on the positive side.

5.2 Airglow Enhancement

Airglow in the F region of the ionosphere is ascribed to the emissions of atomic oxygen in the excited states. The minimum electron energies to excite atomic oxygen which produces airglow at 630 and 557.7 nm are 3.1 and 5.4 eV, respectively. Airglow enhancements have been observed in the O-mode HF heating experiments.

Parametrically excited electron plasma waves and upper hybrid waves can accelerate electrons to suprathermal level; the produced electron flux in the energy range of ~3 to 6 eV excites atomic oxygen through inelastic collisions to produce airglow, which consists of emissions mainly at 630 and 557.7 nm. The enhancement of airglow at 777.4 nm was also observed when the HF transmitter was operated at frequencies near the second and third harmonic of the electron cyclotron frequency. The minimum electron energy to excite 777.4 nm emissions is 10.7 eV, which is well beyond the suprathermal level. Electron harmonic cyclotron resonance interaction is the mechanism of producing such high energy level electron flux.

5.2.1 *Harmonic cyclotron resonance interaction*

The trajectory of the electron, in a wave field $\mathbf{E}(\mathbf{r}, \mathrm{t})$, is governed by the equations

$$\frac{d\mathbf{r}(t)}{dt} = \mathbf{v}(t) \tag{5.1}$$

and

$$\frac{d\mathbf{v}(t)}{dt} = -\frac{e}{m_e}\mathbf{E}(\mathbf{r}, t) - \Omega_c \mathbf{v}(t) \times \hat{\mathbf{z}} \tag{5.2}$$

where $\Omega_c = |\Omega_e|$ is the electron cyclotron frequency and $\hat{\mathbf{z}}$ is set in the magnetic field direction, i.e., $\mathbf{B}_0 = \hat{\mathbf{z}}\,B_0$. Consider an ES wave field $\mathbf{E}(\mathbf{r}, t) = \int[(\hat{\mathbf{x}}k_x + \hat{\mathbf{z}}k_z)/k]E_k \cos(k_x x + k_z z - \omega t)dk/2\pi$, and set $v^- = v_x - iv_y; E_{kx} = k_x E_k/k$ and $E_{kz} = k_z E_k/k$, the trajectory equations are integrated to be

$$\mathbf{r}(t) = \mathbf{r}_0(t) - \int_0^t ds \overleftrightarrow{\mathbf{H}}(t-s) \cdot \frac{\mathbf{E}[\mathbf{r}(s), s]}{B_0} \tag{5.3}$$

$$\begin{aligned} v^- &= e^{-i\Omega_c t}\left\{v_{\perp 0}e^{-i\theta_0} - \frac{e}{m_e}\int \frac{dk}{2\pi}E_{kx}\right. \\ &\quad \left.\times \int_0^t ds e^{i\Omega_c s}\cos[k_x x(s) + k_z z(s) - \omega s]\right\} \\ &= e^{-i\Omega_c t}\left\{v_{\perp 0}e^{-i\theta_0} - \frac{e}{m_e}\int \frac{dk}{2\pi}E_{kx}\int_0^t ds\ e^{i(\omega+\Omega_c)s}e^{-i[k_x x(s)+k_z z(s)]}\right. \\ &\quad \left. + e^{-i(\omega-\Omega_c)s}e^{i[k_x x(s)+k_z z(s)]}\right\} \end{aligned} \tag{5.4}$$

and

$$v_z = v_{z0} - \frac{e}{m_e}\int \frac{dk}{2\pi}E_{kz}\int_0^t ds \cos[k_x x(s) + k_z z(s) - \omega s] \tag{5.5}$$

where $\mathbf{r}_0(t) = \overleftrightarrow{\mathbf{H}}(t)\cdot\mathbf{v}_0/\Omega_c$, $\mathbf{v}_0 = \mathbf{v}(0) = v_{\perp 0}(\hat{\mathbf{x}}\cos\theta_0 + \hat{\mathbf{y}}\sin\theta_0) + \hat{\mathbf{z}}\, v_{z0}$, and $\overleftrightarrow{\mathbf{H}}(t) = (\hat{\mathbf{x}}\hat{\mathbf{x}}+\hat{\mathbf{y}}\hat{\mathbf{y}})\sin\Omega_c t - (\hat{\mathbf{x}}\hat{\mathbf{y}}-\hat{\mathbf{y}}\hat{\mathbf{x}})(1-\cos\Omega_c t) + \hat{\mathbf{z}}\hat{\mathbf{z}}\Omega_c t = -\overleftrightarrow{\mathbf{N}}(t)$ (given in (3.6)).

Equation (5.4) shows that electron gyrates, according to the right-hand rule, around the magnetic field at a cyclotron frequency $f_c = \Omega_c/2\pi$. When it interacts with a left-hand/right-hand (with respect to the magnetic field direction, set along the z axis) circularly polarized wave at a frequency f, the rotation frequency of the wave field seen by the electron is shifted to $f_1 = f \pm f_c - k_z v_{z0}/2\pi$, where the Doppler shift term, $k_z v_{z0}/2\pi$, is attributed to the electron motion along the magnetic field. Thus a right-hand circularly polarized wave field at a frequency $f = f_c + k_z v_{z0}/2\pi$ becomes a dc field as seen by the electron. This is the (fundamental) cyclotron resonance, and the interaction leads to nonzero average in the electron speed.

5.2.2 *Finite Larmour radius effect*

If $f - k_z v_{z0}/2\pi = nf_c$, where the integer $n > 1$ and the wave field is uniform in space, then the net effect on the electron velocity is zero because electron is interacting with a uniform ac field. If the wave field has a spatial variation, for example in the x-direction, but the Larmour radius of the electron gyration is assumed to be zero, i.e., assuming that the electron interacts with the wave field at a fixed location (at the guiding center) where the wave field is a constant ac field, the net effect of interaction on the electron velocity is still zero.

The finite Larmour radius $R_c = v_\perp/\Omega_c$ of the electron gyration makes the electron to experience the dependence of the wave field on $k_x x$; the field intensity in the acceleration phase and in the deceleration phase are different, it results in a nonzero net effect on the electron speed in one gyration period. This is called "finite Larmour radius effect", which is weighed in terms of $k_x v_\perp/\Omega_c$, the product of the wavenumber k_x (measuring the spatial variation rate of the wave field) and the Larmour radius R_c of the electron gyration (covering the extent of the spatial variation).

5.2.3 *Harmonic cyclotron resonance trajectory*

In the harmonic cyclotron resonance situation, $f - k_z v_{z0}/2\pi - f_c \sim Nf_c$, where the integer $N \geq 1$, the x coordinate of the electron resonant trajectory is obtained to be

$$x(t) = \frac{v(t)}{\Omega_c}\sin\theta(t) - \frac{v_{\perp 0}}{\Omega_c}\sin\theta_0 \tag{5.6}$$

where $\theta(t) = \theta_0 + \Omega_c t + \varphi(t)$; $v_\perp(t) = [(v_{\perp 0} + \beta)^2 + \alpha^2]^{1/2}$ and α and β are determined self-consistently to be

$$\alpha = \frac{e}{2m_e}\int \frac{dk}{2\pi} E_{kx} \int_0^t ds[J_{n-1}(w')\sin\Theta'_- - J_{n+1}(w')\sin\Theta'_+] \tag{5.7}$$

$$\beta = -\frac{e}{2m_e}\int \frac{dk}{2\pi} E_{kx} \int_0^t ds[J_{n-1}(w')\cos\Theta'_- + J_{n+1}(w')\cos\Theta'_+] \tag{5.8}$$

In (5.7) and (5.8), $J_p(w')$ is the Bessel function of order p, $w = k_x v_\perp / \Omega_c$, and $w' = w(s)$; $\Theta'_\pm = \Phi + (n \pm 1)\varphi' + k_z(z' - v_{z1}s)$, $\Phi = -\eta \sin\theta_0 + n\theta_0$, $\eta = k_x v_{\perp 0}/\Omega_c$, $\varphi' = \varphi(s)$, $z' = z(s)$, and $\varphi(t) = \tan^{-1}[\alpha/(v_{\perp 0} + \beta)]$. Because $J_q(0) = 0$ for $q \geq 1, \alpha = \beta = 0$ if the finite Larmour radius effect is neglected, i.e., if $w = 0$ is assumed. Without the finite Larmour radius effect, $v_\perp(t) = v_{\perp 0}$, there is no net acceleration from harmonic cyclotron resonance interaction.

Substituting (5.6) into (5.4) and (5.5) yields,

$$v^- = v_\perp(t) e^{-i\theta(t)} \tag{5.9}$$

$$v_z = v_{z0} - \frac{e}{m_e} \int \frac{dk}{2\pi} E_{kz} J_{n-1}(\eta) \times \int_0^t ds J_n(w') \cos[\Phi + n\varphi' + k_z(z' - v_{z1}s)] \tag{5.10}$$

and

$$z = v_{z0}t - \frac{e}{m_e} \int \frac{dk}{2\pi} E_{kz}\, J_{n-1}(\eta) \times \int_0^t ds(t - s) J_n(w') \cos[\Phi + n\varphi' + k_z(z' - v_{z1}s)] \tag{5.11}$$

As shown by (5.7) and (5.8), the finite Larmour radius effect works to shift down the wave frequency to the fundamental cyclotron resonance frequency. The fractional dc field, $\propto J_{n-1}(w)E_{kx}$, experienced by the resonance electron drives the acceleration. The energizing process is also enhancing the finite Larmour radius effect, $\propto J_{n-1}(w)$ which increases with $w(= k_x v_\perp/\Omega_c)$, to provide a positive feedback to the interaction. Giving sufficient large $w(0)$, i.e., initial Larmour radius effect, the harmonic cyclotron resonance is an effective process to produce energetic electrons. On the other hand, there is no feedback in the fundamental cyclotron resonance interaction.

5.2.4 *Numerical results*

Thus, heating at the harmonic cyclotron resonances directly by the HF heater wave is not effective because the wavenumber of the HF heater wave is very small. On the other hand, an O-mode HF heater wave can excite parametric instabilities to produce short-scale upper hybrid waves, which can interact effectively with electrons at cyclotron harmonic resonance. This is demonstrated in Fig. 5.2 showing that electrons are accelerated by parametrically excited upper hybrid waves via cyclotron harmonic resonance interaction to energetic levels. Extra-thermal electrons enhance airglow and produce ionizations when colliding with the background neutral particles.

5.3 Energetic Electron Flux

Energetic electron fluxes in the energy range from 10 to 25 eV have been detected in situ by a probe in rocket as well as on the ground by radar inferred by ultraupshifted frequency band and by the enhancement of airglow at 777.4 nm during the O-mode HF heating.

Plasma waves excited by the parametric instabilities accelerate electrons to such high energy level. Langmuir waves can resonantly interact with electrons through the Doppler shift $\mathrm{f} - \mathrm{k_z v_{z0}}/2\pi \sim 0$; upper hybrid waves can resonantly interact with electrons through the harmonic cyclotron shift, which generates energetic electron flux as shown in Fig. 5.2.

5.4 Artificial Ionospheric Layers (AILs)

Observations of optical emissions enhanced by the O-mode HF heater wave at frequency set near double the electron gyro-frequency ($\sim$2.85 MHz) indicated new ionized layers were produced. Experiments were conducted during twilight and early evening hours in Alaska local time, when the photo-ionization was weak. The wave–ionosphere interaction occurs in the region around 230 to 250 km, where the O^+ ions are dominant The ionization energy of the atomic oxygen

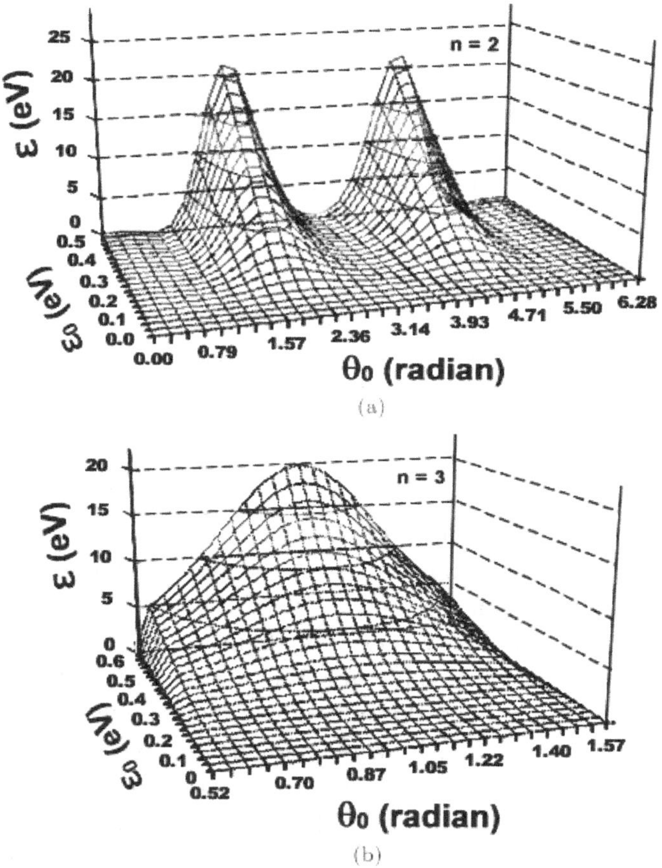

Fig. 5.2. The dependencies of the steady state electron energy ε on the initial energy ε_0 and the azimuthal angle θ_0 of the initial transverse velocity (in the cylindrical coordinates) after acceleration by upper hybrid waves having a central wavelength of 0.5 m and (a) a field intensity of 0.015 V/m through second harmonic resonance interaction and (b) a field intensity of 0.025 V/m through third harmonic resonance interaction.

"O" is 13.6 eV. The experimental results showed that the enhanced optical emissions descended in the background F-region ionosphere and a relatively thin artificially ionized layer emerged from the base of the ambient F region as seen directly in the ionograms of the digisonde.

AILs were also observed in later experiments with the heater wave frequencies set at 4.34 MHz and at 5.8 MHz, which are around the third and fourth harmonic of the electron gyro-frequency. Digisonde ionograms show that AILs emerge from the base of the ambient F region, similar to those formed with the heater wave frequency set at second harmonic of the electron gyro-frequency.

5.4.1 *Likely courses*

In the parametric excitation, upper hybrid wave is excited by the HF heater wave of frequency f_0 through coupling with lower hybrid wave of frequency f_{Lks}. Thus, the upper hybrid wave frequency $f_{uk} = f_0 - f_{Lks}$, where as the lower hybrid frequency $f_{Lks} = f_{LH}\xi^{1/2}$ with $f_{LH} \sim (\Omega_c\Omega_i)^{1/2}/2\pi$ and $\xi = 1 + (m_i/m_e)(k_z^2/k_\perp^2)$. In HAARP, the lower hybrid resonance frequency $f_{LH} \sim 8.3$ kHz. For $\xi = 1$ to 10, $f_{Lks} \sim 8.3$ to 26.3 kHz. Hence, f_{uk} has a bandwidth of 18 kHz, which covers the change of the second harmonic electron gyro-frequency over an altitude range of about 12 km (because the geomagnetic field intensity increases slightly with a decrease of the altitude). It makes accessible for the match of second harmonic cyclotron resonance over this altitude region. Thus, electrons can be accelerated continuously while moving downward through second harmonic cyclotron resonance interaction with spatially and frequency distributed upper hybrid waves along a slightly increasing geomagnetic field. The density of energized electrons which cause optical emissions increases with the decrease of the altitude; at the bottom of the resonance region, the energetic electron flux reaches the maximum intensity and the energetic electrons in the flux also reach the highest energy level. Those electrons with energy exceeding the ionization energy, 13.6 eV, of the atomic oxygen "O" cause a major ionization at the bottom of the F region where the neutral density is high. This explains the descending feature of the enhanced optical emissions in the development of an AIL and the emergence of the AIL as a relatively thin layer at the bottom of the F region.

Moreover, when parametrically excited Langmuir waves near the HF reflection height evolve nonlinearly to become turbulent, these waves are also effective in generating suprathermal electrons to create an artificial ionized layer (AIL), which in turn lowers the HF reflection. In the self-consistent interaction process, the created ionized layer is descending. The descent process stops at the bottom of the F region where ionization, recombination, and diffusion loss reach equilibrium. Because this generation mechanism does not require HF frequencies to be near electron gyroharmonics, it needs further experimental validation.

Although Langmuir PDI and upper hybrid PDI are competing in the HF excitation (discussed in Sec. 5.7.3), both waves are generated concurrently. Therefore, these two likely courses are additive to each other.

5.5 Artificial Spread-F

Digisonde is an HF radar probing the electron density distribution in the bottomside of the ionosphere. This radar transmits O-mode and X-mode sounding pulses with the carrier frequency f swept from 1 to 10 MHz and records sounding echoes in an ionogram. A sounding echo represents a backscatter signal of a sounding pulse from a layer of the ionosphere. Because digisonde radiates at a large cone angle, each sounding pulse can be decomposed into rays having different incident angles; thus, these rays propagate at different ray trajectories. Only backscattered rays can return to the digisonde and are recorded as the ionogram echoes. In the unperturbed ionosphere, only rays close to vertical transmission are backscattered from a layer where the electron plasma density $N(h')$ matches the cutoff density $N_{cO} = (f/9000)^2$ cm^{-3} or $N_{cX} = f(f - f_c)/(9000)^2$ cm^{-3} set by the carrier frequency f of the O-mode or X-mode sounding pulse. The virtual height $h'(f) = c\tau/2$ of an echo is determined by the time delay τ of the echo, where c is the speed of light in free space. The O-mode virtual height trace $h'(f)$ of the sounding echoes in the ionogram is converted to a true height trace $h(f)$ by the plasma density-dependent group velocity determined self-consistently; $h(f)$ is then inverted to obtain the plasma density profile $N(h)$.

5.5.1 *Spread of the virtual height traces in an ionogram*

When rays propagate in a perturbed ionosphere, ray intensity is attenuated by the scattering of the ionospheric perturbation distributed along its unperturbed ray trajectory. A series of small angle scatterings is essentially equivalent to split each ray into sub-rays which propagate at different trajectories deviated from the unperturbed one. Hence, echoes from unperturbed backscattering trajectories (for vertically incident rays) will be weakened by the background perturbations, but some of the sub-rays of obliquely incident rays could have backscattered trajectories to become sounding echoes. In other words, in the presence of large-scale field-aligned density irregularities (FAIs with scale lengths of a few hundreds of meters), the echoes in the ionogram are contributed by the returns of vertically as well as obliquely incident rays. These echoes at each radar frequency have different return times; it results in the spread of the virtual height traces in the ionogram. In naturally perturbed ionosphere, the spread usually appears in the frequency band corresponding to the F region of the ionosphere and is thus termed "Spread-F".

5.5.2 *Experimental observations*

In HF heating experiments, spread-F induced by the O-mode HF heater wave has been observed and is termed "artificial spread-F". This is illustrated in Fig. 5.3 comparing a pair of ionograms without and with heating effect recorded during the heating experiment conducted on November 20, 2009 from 21:10 to 23:00 UTC (12:10 to 14:00 local time).

As shown, the artificial spread-F extends from 2.1 to 3.5 MHz, indicated by arrows; on the other hand, the parametric instabilities excited by the O-mode heater of 3.2 MHz occur only locally near the HF reflection height (at 3.2 MHz) and around upper hybrid resonance region (at 2.88 MHz).

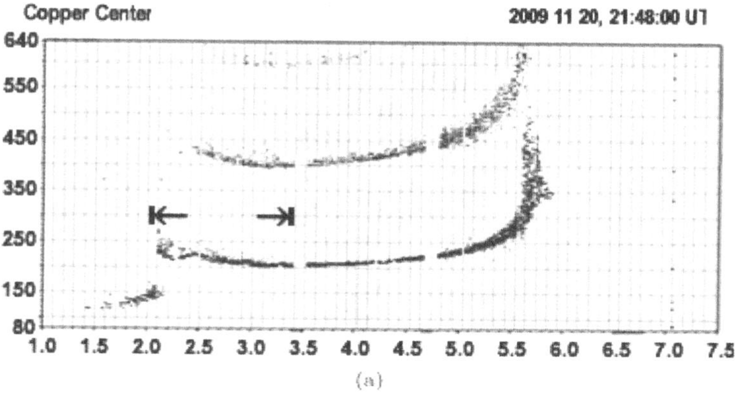

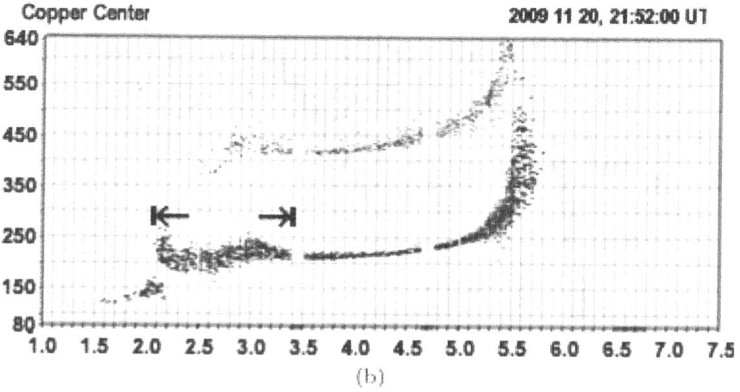

Fig. 5.3. A pair of ionograms (a) without heating effect and (b) with heating effect; the ionogram in (b) was acquired beginning the moment the heater turned off after the 3.2 MHz O-mode heater turned on for 2 minutes. The arrows highlight the region of comparison.

Ray tracing introduced in Chapter 4 is illustrated in the following two sections to justify the discussion presented in Sec. 5.5.1.

5.5.3 *Trajectories of vertically incident rays in the absence/presence of FAIs*

The ray trajectory $\mathbf{r}(t)$ of a mode is governed by (4.14) and (4.15). A cartoon of large-scale FAIs appearing in the F region of the ionosphere is presented in Fig. 5.4(a); x and z axes are in the

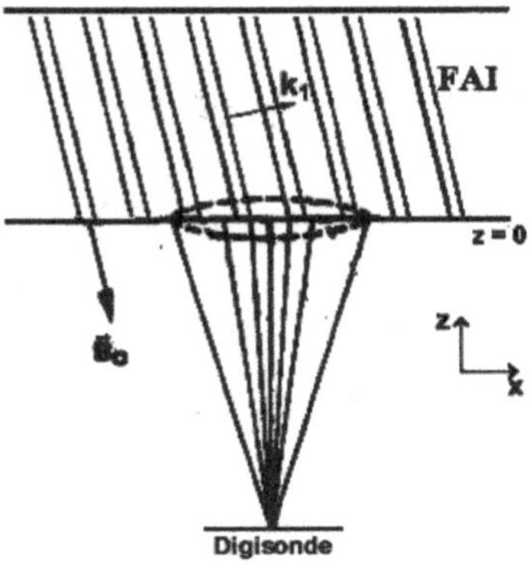
FAI
k1
z = 0
z
x
Digisonde

(a)

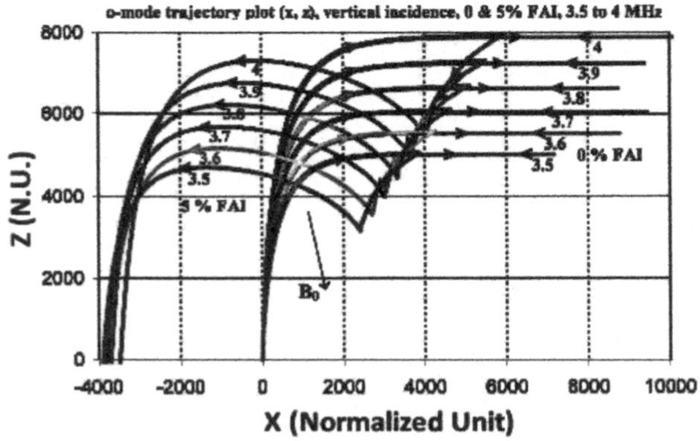
o-mode trajectory plot (x, z), vertical incidence, 0 & 5% FAI, 3.5 to 4 MHz
8000
6000
4000
2000
0
-4000
-2000
0
2000
4000
6000
8000
10000
Z (N.U.)
X (Normalized Unit)
4
3.9
3.8
3.7
3.6
3.5
5 % FAI
0 % FAI
B0

(b)

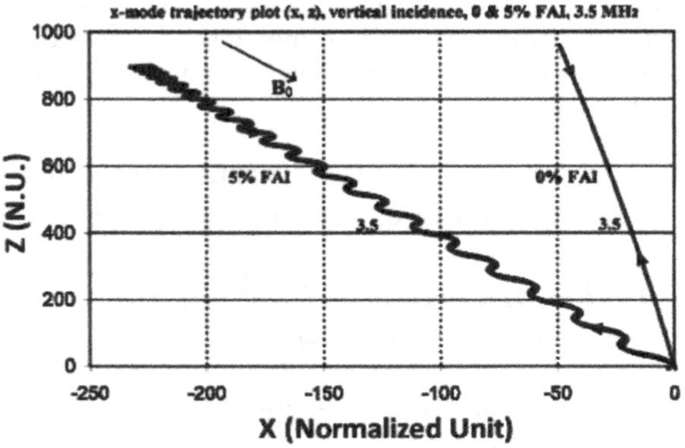
x-mode trajectory plot (x, z), vertical incidence, 0 & 5% FAI, 3.5 MHz
1000
800
600
400
200
0
-250
-200
-150
-100
-50
0
Z (N.U.)
X (Normalized Unit)
B0
5% FAI
0% FAI
3.5
3.5

(c)

←

Fig. 5.4. (*figure on facing page*). (a) Digisonde beam entering ionosphere in the presence of FAIs; (b) and (c) vertically incident ray trajectories in the absence of FAIs and in the presence of 5 % FAIs: (b) O-mode rays of 3.5 to 4 MHz and (c) X-mode ray of 3.5 MHz. The axes x and z in (b) and (c) are normalized to $k_0 = \omega_0/c$. Magnetic dip angle of 76° is adopted.

horizontal and vertical directions and the magnetic dip angle of 76° is adpted. A sounding pulse of the digisonde is decomposed into a bunch of rays as also shown in Fig. 5.4(a); FAIs will modify the trajectories of rays. Only the backscatter signals are recorded in the ionograms. Figure 5.4(b) demonstrates the change of the trajectories of vertically incident O-mode rays in the presence of FAIs, assumed to be at 5% level (i.e., $\Delta N/N_0 = 0.05$, where N_0 is the background plasma density at a reference layer and ΔN is the density variation amplitude of FAIs). In the absence of FAIs, the vertically incident rays of 3.5 to 4 MHz (gray color and labeled with 3.5 to 4) are backscattered. In the presence of FAIs, each vertically incident ray is expected to be split by the FAIs; one part of the ray still propagates along the unperturbed trajectory and is backscattered with lower echo intensity; the other part propagates along modified trajectories, among those a representative one is illustrated in Fig. 5.4(b). As shown, the modified trajectories of vertically incident rays (black color and labeled with 3.5 to 4) do not channel backscatter signals anymore. A similar demonstration for the vertically incident x-mode ray of 3.5 MHz is presented in Fig. 5.4(c). As shown, the vertically incident ray is ducted by the FAIs, the return signal propagates along the geomagnetic field, rather than propagating vertically downward. The x and z axes in Figs. 5.4(b) and 5.4(c) are normalized to $k_0 = 2\pi f/c$, where f is the frequency of the sounding pulse. For example, at 4 MHz, $k_0 = 0.0837758 m^{-1}$. Thus, $1/k_0 = 11.93662$ m. So, the axes for 4 MHz ray are normalized to 11.94 m. It increases to 13.64 m for 3.5 MHz ray.

5.5.4 *Trajectories of backscattered rays in the presence/absence of FAIs*

Therefore, in the presence of FAIs, the traces in the ionogram are not contributed only by the returns of the vertically incident rays, channeled through the unperturbed vertical ray trajectory. The trajectories of backscatter rays in the presence of 5% FAIs for both O- and X-mode sounding signals from 3.5 to 4 MHz are determined. In the ray tracing, FAIs are represented by a single sinusoidal function with a spatial period of 200 m and an amplitude $\Delta N = 0.05\ N_0$, where N_0 is the plasma density at a reference layer, $z = 0$, having a plasma frequency of 2.26 MHz ($3.2\text{MHz}/\sqrt{2}$); the background plasma density has a linear scale length of 30 km. The results are shown in Fig. 5.5, and rays with proper incident angles are backscattered. The backscatter trajectories in the absence of FAIs are also plotted in Fig. 5.5 (light gray color in Fig. 5.5(a) and trajectories in $X < 0$ region in Fig. 5.5(b)) for comparison.

Although large scale FAIs can be generated directly by the HF heater wave through thermal filamentation instability, O-mode heater generates sheet-like FAIs parallel to the meridian plane, which are not effective to enhance virtual height spread of the sounding echo. Another mechanism of large-scale FAI generation is a thermal instability, which is driven by the downward and upward heat flow from localized heat sources under anomalous heating of parametric instabilities excited near the HF reflection region and the upper hybrid resonance region. The generation occurs in an extended region along the geomagnetic field, where artificial spread-F is induced. It explains the observation presented in Fig. 5.3 that the artificial spread-F appears in an extended region.

5.6 Ionization Enhancement

Artificial ionization enhancement was also observed when the O-mode HF heater wave frequency was not close to a harmonic of the electron cyclotron frequency.

Experiments were conducted on November 16 and 20, 2009 around local solar noon when the photo-ionization was strong and

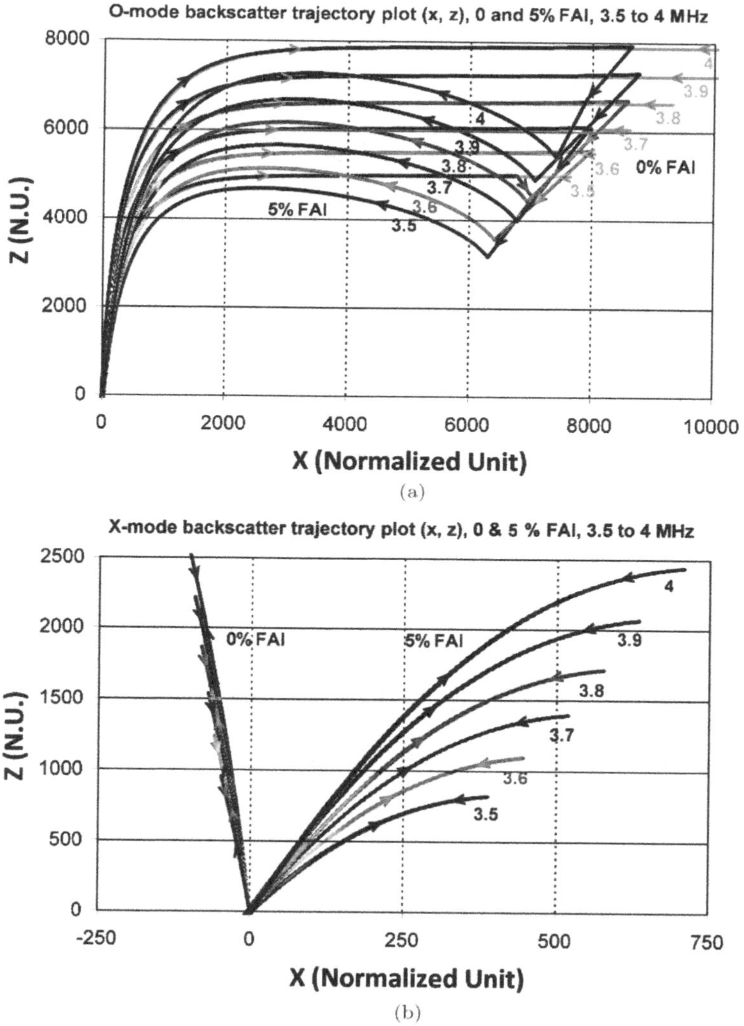

Fig. 5.5. Trajectories of backscatter rays of 3.5 to 4 MHz in the absence of FAI and in the presence of 5% FAIs, (a) O-mode rays and (b) X-mode rays.

the wave–electron interaction occurred in the lower F region (<180 km) of the ionosphere, where the electron–ion effective recombination coefficient depends strongly on the electron temperature T_e. Through parametric instabilities and thermal diffusion, anomalous electron

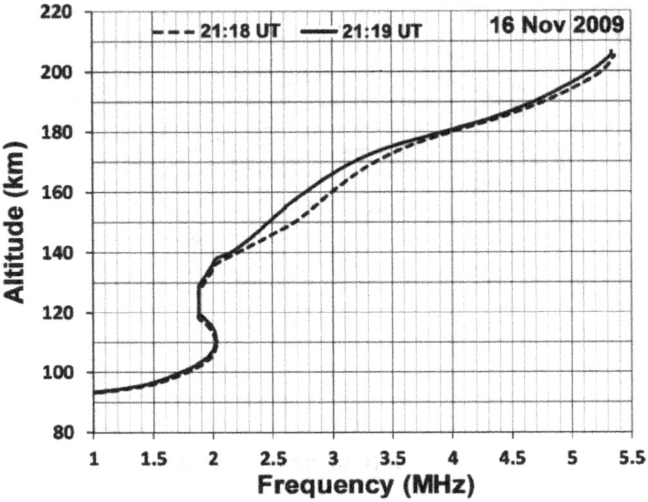

Fig. 5.6. Sounding echo frequency (electron density) vs. height at the two recording times.

heating reduced the recombination coefficient over a large region, which changed the balance between the photo-ionization and the recombination loss, giving rise to an electron density enhancement in the heated region below ~180 km. This is demonstrated in Fig. 5.6, in which sounding echo frequency (i.e., electron density) distributions (a) at 21:18:00 UT, after the heater off within 10 seconds and (b) at 21:19:00 UT, after the heater was off more than 60 seconds, are presented for comparison. As shown, the distribution at 21:18:00 UT has a higher density in the entire modified region. The percentage of the electron density increase exceeds 10% over a large altitude region (>30 km) from below to above the HF reflection height of ~170 km.

5.7 Artificial Cusp

A virtual height bump usually occurs at frequencies near foF1 in the presence of an F1 layer; the true height profile indicates that there is a density ledge (cusp) at foF1, which retards the propagation of sounding signals with frequencies f close to foF1. Spread-F normally extends over a frequency band in the ionogram and is recognized to be caused by the large-scale FAIs in the plasma.

5.7.1 *Bump in the ionogram trace*

However, localized anomalous echo spread appearing as a bump in the ionogram trace was also observed in the O-mode HF heating experiments. This is illustrated in Fig. 5.7, showing two heater off ionograms acquired at 21:42 UT and 21:43 UT after the O-mode heater wave of 3.2 MHz was turned on at 21:40 UT for 2 minutes.

5.7.2 *Digisonde exploring upper hybrid instability*

Both ionograms show spread of echoes, except the spread in 21:42 UT is considerably larger and contain a noticeable bump located next to the plasma frequency (~2.88 MHz) of the upper hybrid resonance layer of the HF heater wave. This heating-induced bump in the ionogram trace at 2.88 MHz is similar in appearance to the virtual height bump at foF1 in the presence of a F1 layer; the similarity suggests that there is a heater-stimulated ionization ledge (cusp) in the upper hybrid resonance region, which explains the appearance of the bump. This ledge can be created by the thermal pressure force and the ponderomotive force exerted by the plasma waves of the parametric instabilities excited by the HF heater wave in the upper hybrid resonance region.

Due to the field-aligned nature of the upper hybrid waves, these waves could not be detected directly by the UHF/VHF radars. The distinctive virtual height bump in Fig. 5.7 suggests that the digisonde can be a key diagnostic instrument to explore the upper hybrid waves excited in the HF heating experiment.

5.7.3 *Competition between Langmuir parametric decay instability and upper hybrid parametric decay instability*

The time development of bumps in the ionogram trace, manifesting a competition of different parametric instabilities, was also observed.

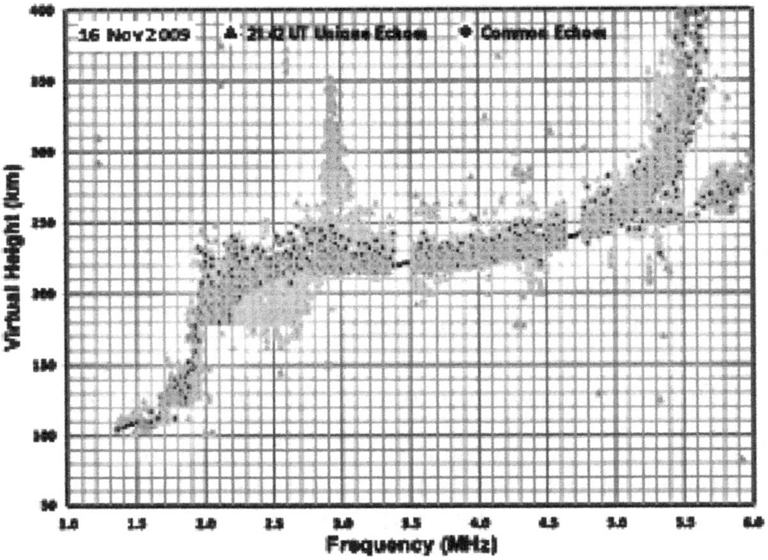

(a)

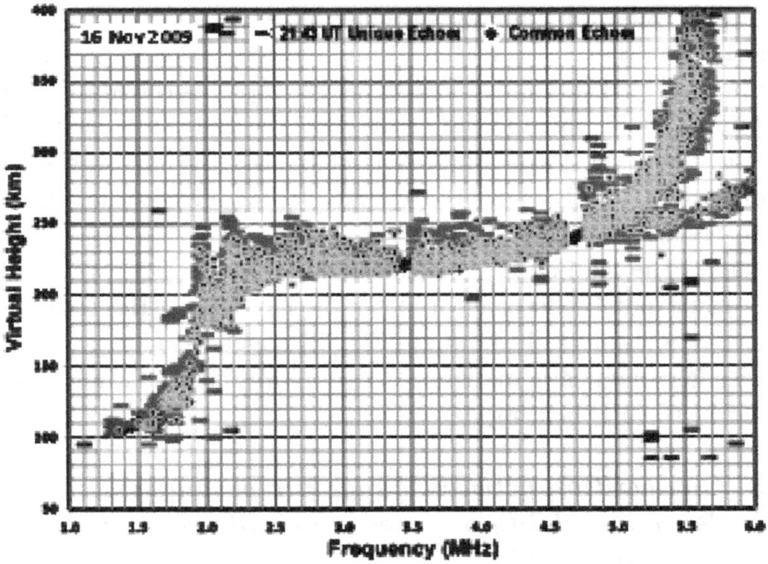

(b)

Fig. 5.7. The echoes (unique and common) extracted from the (a) 21:42 UT ionogram and (b) 21:43 UT ionogram.

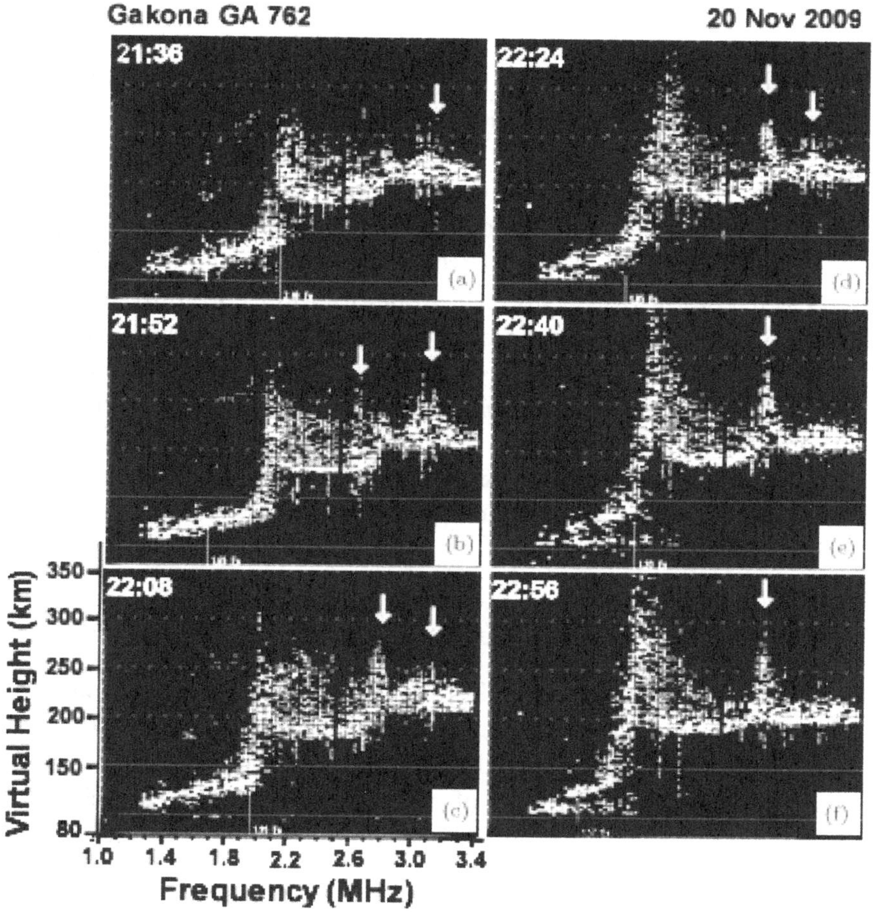

Fig. 5.8. Sequence of ionograms (a–f) showing the evolution of the virtual height spread. Arrows locate the likely bumps in the virtual height spread.

This is demonstrated in a sequence of six ionograms, presented in Figs. 5.8(a–f). The echoes in each of the ionograms were acquired beginning the moment the O mode heater wave turns off. The results show the time change of the virtual height spread as well as the development of the bumps next to the HF reflection height and next to the upper hybrid resonance layer. In Fig. 5.8(a), one bump indicated by an arrow appears in the region below 3.2 MHz (i.e., HF reflection height). Based on the location of this speculative bump, one may infer that the Langmuir PDI was excited by the HF heater wave. As the experiment proceeded, Fig. 5.8(b) shows that a second bump indicated by an arrow appears in the region below 2.88 MHz (i.e., upper hybrid resonance height) also appears. One may infer that the Langmuir PDI and upper hybrid PDI were excited concurrently by the HF heater wave. The Langmuir PDI bump shifted down slightly in frequency and was gradually weakened as indicated in Figs. 5.8(c) and 5.8(d); on the other hand, the upper hybrid PDI bump shifted up slightly in frequency and became stronger. This upper hybrid PDI bump evolved to a steady state level in the subsequent ionograms shown in Figs. 5.8(e) and 5.8(f), and the Langmuir PDI bump disappeared completely. This development infers that the PDI was suppressed by instabilities excited in the upper hybrid resonance region, which drain heater energy before heater reaches the PDI excitation region.

5.8 Wideband Absorption

Anomalous attenuation of HF radar echoes over a wideband was observed in Boulder heating experiments; it was attributed to large angle Bragg scattering of HF radar signals by HF-induced (via parametric instabilities) FAIs of a few tens of meters; the scattered signals did not return to the radar receiver to cause a loss in echoes′ intensities.

5.8.1 *Observations*

Wideband absorption was also observed in HAARP heating experiments conducted on July 27, 2011. In Fig. 5.9, ionograms presented in Columns A and D were recorded during the heater on periods and these in Columns B and C were recorded during the heater off periods.

The 10 heater-on ionograms were recorded after turning on the heater for 30 second, and the 10 heater-off ionograms were recorded right after turning off the heater. Heating leads to the decrease of

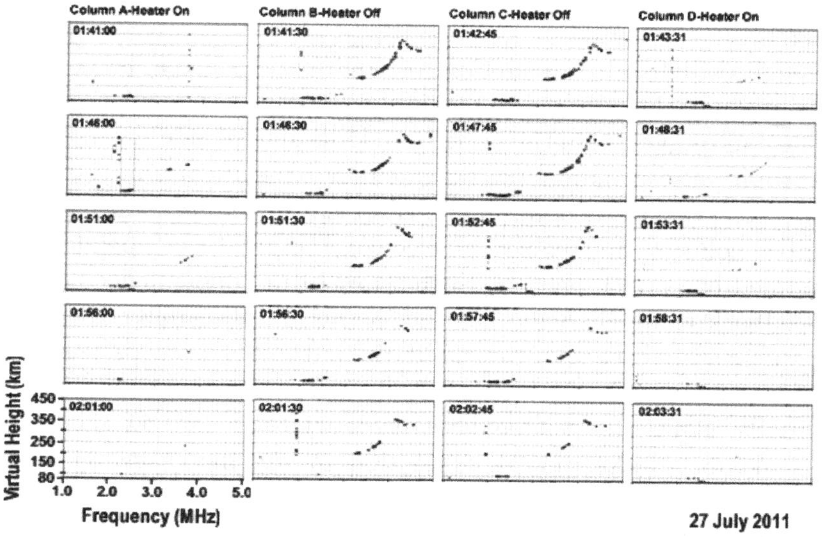

Fig. 5.9. Disappearance of the ionogram echoes over a wideband while the heating experiment, conducted on July 27, 2011, was progressing, suggesting development of heater-induced density irregularities; Columns A and D show heater-on ionograms and Columns B and C show heater-off ionograms.

the ionogram echo intensity. Adjusting the threshold level of the digisonde receiver to record only the strongest echoes, the heating effect can be manifested by the disappearance of ionogram echoes in comparing a pair of the heater-on and heater-off ionograms. In fact, the heater-on ionograms recorded 3 minutes after the start (at 01:38:00) of the experiment (i.e., at 01:41:00 and later) already show disappearance of echoes.

Moving down along each column of Fig. 5.9, one observes clearly a decrease in the number of ionogram echoes in both on and off ionograms. It shows that the heating effect is accumulative, and the heater-off period is not long enough to damping away heating induced perturbations. This suggests that the perturbations are medium-scale FAIs, which have such long damping times. The ionogram echoes disappear almost completely in the on ionograms of the last two rows (i.e., from 01:56 to 02:01 UT). This represents wideband absorption on the HF signals.

5.8.2 *Physical mechanisms*

Though wideband attenuation of ionosonde signals is generally attributed to the presence of medium-scale density irregularities (i.e., a few tens of meters) in the background ionosphere, nonlinear thermal instability also produces periodic density irregularities in the D and lower E regions of the ionosphere. Both elastic and inelastic electron–neutral collision frequencies are electron temperature dependent and modified by the HF heating. Such temperature dependence provides a feedback channel to the heating; a thermal instability is excited. As the heating exceeds a threshold, the induced density perturbation evolves nonlinearly to form periodic density irregularities varying spatially along the geomagnetic field, which is 14° off the zenith in Gakona, Alaska. The combination of the increasing D region absorption, which weakens the sounding signals, and the E region scattering by the produced periodic density irregularities could also cause the disappearance of ionogram echoes over a wideband.

5.9 Short-scale (Meters) Field-aligned Density Irregularities for Bragg Scattering

Parametrically excited high-frequency plasma waves set up a ponderomotive force acting on electron plasma. This force modifies the plasma density distribution in which ions follow electrons via the induced self-consistent electric field. The induced density perturbation tends to develop in a field-aligned nature because the diffusion loss along the geomagnetic field is large. The density perturbation redistributes the high-frequency plasma waves, which are cumulative in the density depletion regions where the wave phase speed is lower. This in turn reinforces the ponderomotive force to further increase the density perturbation. Such a positive feedback excites an ES filamentation instability, which generates meter-scale FAIs and filaments the intensity of the high-frequency plasma waves. The short-scale FAIs can also be generated directly by the HF heater wave via upper hybrid OTSI, which decays the HF heater wave to two upper hybrid sidebands and a short-scale FAI. It was demonstrated that short-scale density irregularities could scatter ground-transmitted VHF signals (wavelengths of 1 to 10 m) back to the ground at a different site, over the horizon distance away from the transmitter site, a possible communications link.

5.10 Stimulated Electromagnetic Emissions (SEEs)

Short-scale FAIs (0, $\mathbf{k}_I$) can also scatter (convert) high-frequency plasma waves (ω, $\mathbf{k}$) into electromagnetic (EM) radiation (ω_0, $\mathbf{k}_0$) where the frequency ω_0 and wavevector $\mathbf{k}_0$ of the EM radiation are imposed by the frequency and wavevector matching conditions $\omega_0 = \omega$ and $\mathbf{k}_0 = \mathbf{k} \pm \mathbf{k}_I$. Because the wavenumber k_0 of the EM radiation is much smaller than that of electron plasma wave, i.e., $k_0 \ll k$, the wavevector matching condition requires that $\mathbf{k} \cong \mp\mathbf{k}_I$. Thus, the scattered high-frequency plasma wave has to be a near field-aligned mode having a wavenumber close to the wavenumber of an FAI. Emissions with frequency downshifted spectra have been detected on the ground by HF receivers, mainly at high latitude heating sites. The spectra resembled those of upper hybrid/electron Bernstein waves excited by parametric instabilities. These parametrically excited ES waves, which could not be detected by backscatter radar due to their propagation direction being nearly perpendicular to the geomagnetic field, were converted, via scattering by short-scale FAIs, into EM emissions, which propagate downward to the ground. These emissions were generated through instabilities, thus termed "SEEs". SEEs require the presence of short-scale density irregularities and electron waves, which are sensitive to the heater frequency in reference to the electron cyclotron harmonic resonance frequencies. Therefore, various spectral features of SEEs have been observed by fine-tuning of the heater frequency around each electron cyclotron harmonic resonance frequency.

5.11 Plasma Turbulence

In ionospheric HF heating, plasma maintains connection with an external source (HF heater) and a sink of energy and may reach a turbulent state. In the F region of the ionosphere, collision frequency is small, and a major part of the energy flow goes to the excitation of (ES) plasma waves, which broaden spectra by cascading to longer wavelength modes (such as Langmuir condensate modes). This process evolves plasma to a weak turbulent state, where each spectral line is dictated by the dispersion effect. In the kinetic formulation, linear perturbation expansion of the distribution function is employed. However, each perturbation term contains a singularity at the electron–wave resonance velocity, e.g., the RHS of (4.29) at $v = \omega/k$, the phase velocity of the wave in unmagnetized plasma. As the spectral intensity is intensified, the nonlinear effect enhances mode–mode coupling, which further broadens the spectral distribution and makes it difficult to truncate the perturbation expansion.

A renormalization theory was introduced to address the strong turbulence state. In essence, the theory re-sums the singular perturbation expansion terms into a single secular term, which exhibits resonance broadening, i.e., the bandwidth of the resonance velocity is increased from a point to a finite velocity band around ω/k so that the singularity at ω/k is removed.

A strong Langmuir turbulence theory was also developed and applied for electron energization. A large amplitude Langmuir wave is trapped in the self-induced potential well and evolves nonlinearly to form cavities (i.e., cavitons) in the plasma. The depth of each cavity is proportional to the intensity square of the trapped waves. If the dispersive effect is not able to balance the nonlinear effect, the field intensity evolves to very high level, which eventually collapses the caviton. When this occurs, trapped electrons are released and some of those are accelerated to become energetic, i.e., converting the potential energy in the well into kinetic energy. However, the quasi-linear theory indicates that the electron effective temperature T_{eff} in (4.43) also increases with the intensity square of the trapped waves; thus, the dispersion effect increases concurrently with the nonlinear effect. In other words, when the electron quiver speed becomes comparable to the electron thermal speed, the strong Langmuir turbulence theory is not self-consistent with the assumptions employed in the formulation.

Therefore, the energetic electrons are likely generated by the Langmuir/upper hybrid condensate modes, which spread phase velocities to the larger velocity regime. The harmonic cyclotron shift further broadens the resonance velocity band. When the resonance interaction velocity band of each Langmuir/upper hybrid cascade line is broadened by the resonance broadening effect, electrons could be accelerated continuously to reach an energetic level.

We next discuss the mechanism and give the formulation of parametric instabilities in Chapter 6. The analysis of parametric instabilities which are excited by the O-mode HF heater waves transmitted from ground HF heating facilities is presented in Chapter 7.

Problems

P.5.1. In Arecibo, Puerto Rico, the UHF backscatter radar is operated at 430 MHz; what are the wavelengths and propagation directions of the Langmuir waves recorded as HFPLs?

P.5.2. In Prob. 1, the HF heater frequency is 5.1 MHz. If the plasma waves are excited by OTSI, what are the frequencies of the radar echoes?

P.5.3. In Problem 2, what is the plasma frequency of the layer where the backscattering of radar signal occurs?

P.5.4. Consider the example "(1) Ray trajectory in the bottomside of the ionosphere" in Chapter 4; let $\omega_{p0}/2\pi = 5$ MHz and $z_0 = 250$ km, what is the virtual height of the reflection Layer?

P.5.5. In high-latitude ionosphere, the background geomagnetic field B_0 is about 0.5 Gauss; upper hybrid wave is generated for second harmonic cyclotron resonance heating of the ionosphere.

(1) What is the frequency of the upper hybrid wave to be generated?
(2) What is the plasma density of the layer where the upper hybrid wave is generated?

P.5.6. The magnetic dip angle at Gakona, Alaska, (the location of the HAARP heating facility) is 76°. A digisonde transmits O-mode sounding pulses with the carrier frequency f swept from 3 to 7 MHz and records sounding echoes in an ionogram. If wideband absorption occurs, i.e., no sounding echo appears in the ionogram, what is the range of the scale lengths of the FAIs which are responsible for the wideband absorption?

P.5.7. An antenna is designed to have a directivity, which aims the radiation to a certain direction. Thus, the power density S_D of the radiation at the aimed direction is larger, by a factor G named "antenna gain", than that S_i of an isotropic antenna (i.e., radiating uniformly in all direction) radiated at the same power, i.e., $S_D = GS_i$. In order to achieve the same power density at the aimed direction as that of a directed antenna, the isotropic antenna has to raise the radiation power P_i from P_D of the directed antenna by the same factor G, i.e., $P_i = GP_D$, and P_i is called "ERP" of the directed antenna, where P_D is the actual radiation power and G is the antenna gain.

HAARP HF antenna array radiates at 3.6 MW full power and has antenna gain of 25 dB in the frequency band of 4 to 7.6 MHz and 30 dB in the frequency band of 7.6 to 10 MHz. Calculate the ERP of the HF heater in respective frequency band.

P.5.8. The time average power density of the radiation in free space is given by $S = \langle E^2 \rangle / \eta_0$ W/m^2, where $\langle \cdots \rangle$ represents time average and $\eta_0 = (\mu_0/\varepsilon_0)^{1/2} = 377\Omega$ (i.e., $120\pi\Omega$) is the intrinsic impedance of the free space. Let E_0 be the amplitude of the radiation at a distance r from the antenna; thus, $S(r) = E_0^2/2\eta_0$ for a linearly polarized radiation and $S(r) = E_0^2/\eta_0$ for a circularly polarized radiation.

Since the power density at a distance r from an isotropic antenna is also given by $S_i = P_i/4\pi r^2$, we can set up the equation $\langle E^2 \rangle / \eta_0 = GP_D/4\pi r^2$ to determine the free space field amplitude $E_0(r) = (60GP_D)^{1/2}/r$ V/m for a linearly polarized radiation and $E_0(r) = (30GP_D)^{1/2}/r$ V/m for a circularly polarized radiation, from the radiation of a directed antenna having gain G and radiating at power P_D. In the ionosphere, this field amplitude near the HF reflection height is enhanced by a swelling factor of 4 (assuming that there is no D region absorption, e.g., in the nighttime).

Use the Data given in Problem 5 to calculate the field amplitude E_0 of the HAARP HF heater of 4 MHz at its reflection height of 300 km.

P.5.9. In HAARP heating experiment, O-mode HF heater of 5 MHz is transmitted. Assume that PDI is excited at a layer of $\omega_p =$ 4.8 MHz. The background plasma parameters are $T_e = 1500$ K and $T_i = 1000$ K. Plot the phase velocity distribution of first 10 Langmuir cascade lines propagating along the geomagnetic field.

Chapter 6

Parametric Instabilities

6.1 Physical Mechanism

A parametric amplifier uses three coupled resonant circuits (e.g., LC circuits) to convert frequency from one to another. A nonlinear (variac) capacitor in the circuit provides the coupling (i.e., frequency mixing). The three resonant modes in a parametric amplifier are called source (pump), idler (sideband), and signal (decay mode), which oscillate at ω_0, ω_1, and ω_s to satisfy the frequency matching condition $\omega_0 = \omega_1 + \omega_s$. To apply the mechanism of wave amplification from a circuit to a system, one recognizes that the system has to support at least three branches of oscillations and carry nonlinear properties for wave–wave couplings. Plasma meets the criterion.

Plasma is a nonlinear dielectric medium and supports at least three branches of modes; high- and low-frequency plasma modes oscillate in plasma as thermal fluctuations in the absence of external sources. When a large amplitude high-frequency wave $\mathbf{E}_p(\omega_0, \mathbf{k}_p)$ (either electromagnetic (EM) or electrostatic (ES)) appears in plasma, the parametric coupling sets it up as a pump wave to excite high- and low-frequency plasma waves concurrently. The electric field of the high-frequency pump wave launches a quiver motion $v_{eq}(t)$ in the electron plasma. Electrons and ions in the low-frequency thermal fluctuations oscillate together to maintain quasi-neutrality, i.e., $n_{es}(\omega_s, \mathbf{k}_s) \cong n_{is}(\omega_s, \mathbf{k}_s) = n_s(\omega_s, \mathbf{k}_s)$, which facilitates the buildup of $n_s(\omega_s, \mathbf{k}_s)$ even in a low-intensity fluctuating field. Thus, a high-frequency space charge current density $-en_s v_{eq}$ is produced in the electron plasma. This current drives beat waves $\mathbf{E}(\omega, \mathbf{k})$ and $\mathbf{E}'(\omega', \mathbf{k}')$ with their frequencies and wavevectors satisfying the matching conditions:

$$\omega_0 = \omega + \omega_s^* = \omega' - \omega_s \quad \text{and} \quad \mathbf{k}_p = \mathbf{k} + \mathbf{k}_s = \mathbf{k}' - \mathbf{k}_s \qquad (6.1a)$$

These beat waves, in turn, couple with the pump wave to exert on electrons a ponderomotive (radiation pressure) force, which has the same frequency and wavevector as $n_s(\omega_s, \mathbf{k}_s)$. This is a feedback loop in the parametric coupling. The strength of the coupling depends on the nature of the induced beat waves. The coupling is strong as the beat waves are plasma modes. When the feedback is positive and large enough to overcome linear losses of the coupled waves, the coupling becomes unstable and the coupled waves grow exponentially in the expense of pump wave energy. This is called "parametric instability", by which the pump wave $\mathbf{E}_p(\omega_0, \mathbf{k}_p)$ decays to two sidebands $\mathbf{E}(\omega, \mathbf{k})$ and $\mathbf{E}'(\omega', \mathbf{k}')$ together with a low-frequency decay mode $n_s(\omega_s, \mathbf{k}_s)$. The parametric coupling is imposed by the frequency and wavevector matching conditions (1) as well as a threshold condition on the pump electric field intensity. When the decay mode $n_s(\omega_s, \mathbf{k}_s)$ has a finite oscillation frequency, two sidebands cannot satisfy the same dispersion relation concurrently. The frequency-upshifted sideband $\mathbf{E}'(\omega', \mathbf{k}')$ is off resonant with the plasma, its intensity will be much lower than that of the other sideband $\mathbf{E}(\omega, \mathbf{k})$. Thus, this sideband $\mathbf{E}'(\omega', \mathbf{k}')$ is usually discarded to simplify the formulation to three-wave coupling.

The most effective parametric instabilities excited directly by the HF heater wave are (1) parametric decay instability (PDI) and (2) oscillating two-stream instability (OTSI), in both mid-latitude and high-latitude regions. The sidebands are Langmuir waves and upper hybrid waves; however, the instabilities involving Langmuir waves as sidebands have to compete with those excited in the upper hybrid resonance region, where encounters the ground transmitted HF heater wave first and the upper hybrid waves are the sidebands of the instabilities. The wavenumber k_0 of the heater is much smaller than the wavenumbers of the ES sidebands and decay modes; thus a dipole pump, i.e., $\mathbf{k}_p = 0$, is generally assumed. In the following, three matchings of parametric coupling are illustrated.

6.1.1 *Parametric decay instability*

This is a three-wave coupling denoted by

$$\text{EM Pump } (\omega_0, \mathbf{k}_p = 0) \to (\omega, \mathbf{k}) + (\omega_s^*, \mathbf{k}_s) \tag{6.1b}$$

where $\mathbf{k}_s = -\mathbf{k}; (\omega, \mathbf{k})$ and $(\omega_s, \mathbf{k}_s)$ are Langmuir/upper hybrid sideband and ion acoustic/lower hybrid decay mode.

6.1.2 *Oscillating two-stream instability*

This is a four-wave interaction involving following couplings:

$$\text{EM Pump}(\omega_0, 0) \to \begin{cases} (\omega, \mathbf{k}) + (-i\gamma, \mathbf{k}_s) \\ (\omega, -\mathbf{k}) + (-i\gamma, -\mathbf{k}_s) \end{cases} \tag{6.1c}$$

where again $\mathbf{k}_s = -\mathbf{k}$; $(\omega, \pm\mathbf{k})$ and $(\omega_s = i\gamma, \pm\mathbf{k}_s)$ represent Langmuir/upper hybrid sidebands and purely growing density striation/FAI.

The instability processes usually prefer the excited waves to be the plasma modes. When the frequencies and wavevectors of the coupled waves are deviated from the mode values, the instability threshold increases. For example, the threshold of OTSI is higher than that of PDI because purely growing density striation/FAI are not plasma mode and the frequency of the sidebands is also slightly off from that of Langmuir mode. The frequency and wavevector matching conditions of the two processes, which are satisfied under the dispersion relations of the coupled plasma waves, are illustrated in Fig. 6.1, which adopts the special case that $\mathbf{k}_1$ (and $\mathbf{k}_S$) is parallel to the geomagnetic field to keep the plot simple. In the figure, the dispersion curves of the Langmuir wave (represented by the parabola, which asymptotically approaches to the line labeled by $\omega = \mathrm{kv_{te}}$ and its mirror image) and ion acoustic wave (represented by the straight line labeled by $\omega = \mathrm{kC_s}$ and its mirror image) are plotted on the ω–k plane, a dipole pump is represented by a point $(\omega_0, 0)$ on the vertical

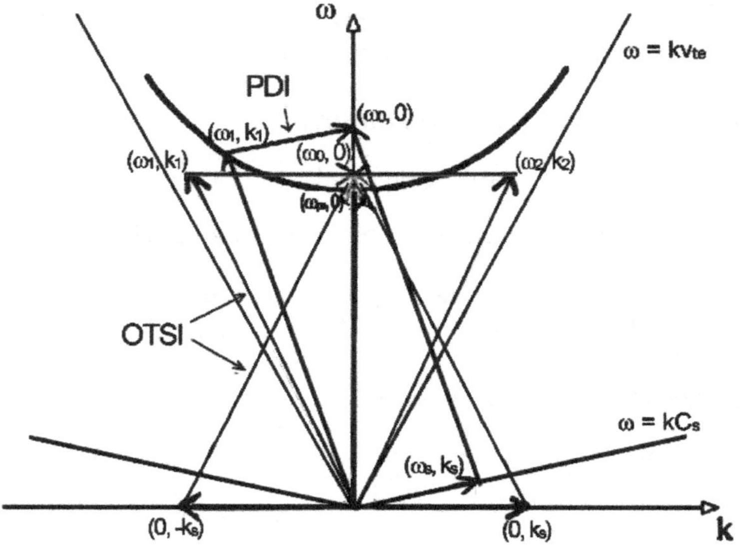

Fig. 6.1. Vector relations showing the frequency and wavevector matching arrangements to identify the plasma modes in the parametric coupling.

axis, and purely growing modes are located on the horizontal axis. Thus, each wave (or mode) labeled by its frequency and wavenumber (ω, k) can be represented by a position vector of the corresponding point on the plane.

In this representation, the frequency and wavenumber matching conditions are combined into a single vector matching condition. Consider PDI matching first. Starting at the point (ω_0, 0), a downward inclined line parallel to the line $\omega = kC_s$ is drawn to intersect with the parabola. The intersecting point (ω_1, k_1) determines the sideband, which is an electron plasma (Langmuir) mode. The difference of the two position vectors (ω_0, 0) and (ω_1, k_1) is then mapped onto the line $\omega = kC_s$ as a position vector (ω_s, k_s). The relationship of the three position vectors shown in the figure illustrates how to identify plasma modes on the dispersion curves to meet the frequency and wavevector matching conditions (6.1a) in the PDI process.

Next, consider OTSI matching. Because the purely growing mode is on the horizontal axis, two symmetric vector matching relations involving two different sidebands, as shown in the figure, can be arranged. Thus OTSI is a four-wave coupling process. Because the real frequencies of the purely growing modes are zero, which is less than the ion acoustic mode frequency, the sidebands' real frequencies also have to be slightly less than the corresponding Langmuir mode frequencies. Using a different point (ω_0, 0) on the vertical axis to represent the pump wave, a horizontal line passing this point is drawn. The sidebands are located slightly outside the parabola at two mirror image points on this line. The exact locations of the sidebands require a detailed analysis of the coupled mode equations. The differences of the position vectors of the pump and sidebands are then mapped onto the k axis to obtain two position vectors $(0, \pm k_s)$, representing the two purely growing modes in the OTSI process.

In HF heating experiments, the wavenumbers of HF-enhanced plasma lines and ion lines (HFPLs and HFILs) are fixed by the wavenumber of backscatter radar. Thus the HFPLs and HFILs contributed by PDI/OTSI originate from narrow regions below/around the O-mode HF reflection height, where the electron plasma frequency $\omega_{pe} = \omega_0$.

6.1.3 *Langmuir (upper hybrid) cascade*

Downward-propagating Langmuir waves excited by OTSI and PDI in the region below the O-mode HF reflection height can become pump waves which propagate to favorable regions (dictated by the matching conditions (6.1a)) to excite new parametric instabilities; those generate frequency-downshifted Langmuir waves to be their sidebands. This is called "Langmuir cascade". Continuous cascade of Langmuir waves through new parametric instabilities broadens the downshifted frequency spectrum of Langmuir waves. A similar description is also applicable to the "upper hybrid cascade", which occurs in the region below the upper hybrid resonance layer where $\omega_{pe}^2 = \omega_0^2 - \Omega_e^2$. However, upper hybrid waves have to propagate along the magnetic field to the favorable regions to excite new parametric instabilities. The permissible number of cascade and the required pump threshold field vary with the plasma, and each cascade process can be distinguished by the nature of the low-frequency decay mode. In the following, a Langmuir cascade process that involves an ion acoustic wave as the decay mode is discussed. This three-wave coupling process is represented by

$$\text{Langmuir Pump } (\omega_1, \mathbf{k}_1) \rightarrow (\omega_2, \mathbf{k}_2) + (\omega_s^*, \mathbf{k}_s) \qquad (6.1d)$$

where $(\omega_2, \mathbf{k}_2)$ and $(\omega_s, \mathbf{k}_s)$ are Langmuir sideband and ion acoustic wave; $\omega_2 = \omega_1 - \omega_s^*$; thus $\text{Re}(\omega_2) \cong \text{Re}(\omega_1)$ and $\mathbf{k}_2 \cong -\mathbf{k}_1$; $\mathbf{k}_s = \mathbf{k}_1 - \mathbf{k}_2 \cong 2\mathbf{k}_1$.

Frequency and wavevector matches in each Langmuir cascade are illustrated in Fig. 6.2, which plots a simple situation that $\mathbf{k}_1$, $\mathbf{k}_2$, and $\mathbf{k}_s$ are parallel to the geomagnetic field. Similar to Fig. 6.1, the dispersion curves of the Langmuir wave and the ion acoustic wave are plotted on the ω–k plane, each wave (or mode) labeled by its

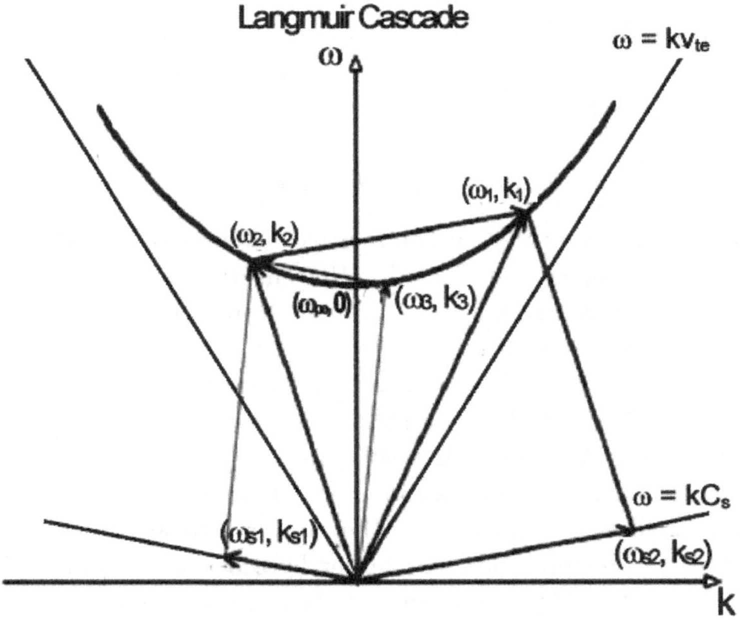

Fig. 6.2. Vector relations showing the frequency and wavevector matching arrangements on the dispersion plane to identify the plasma modes in the parametric coupling.

frequency and wavenumber (ω, k) becomes a point on this ω–k plane and is represented by a position vector of this point.

Starting at a point $(\omega_1, \mathbf{k}_1)$ on the RHS of the parabola, which represents a Langmuir wave excited by PDI, a downward inclined line parallel to the line labeled $\omega = kC_s$ is drawn to intersect with the parabola. The intersecting point $(\omega_2, \mathbf{k}_2)$ on the LHS of the parabola identifies the sideband. The vector between the two points $(\omega_1, \mathbf{k}_1)$ and $(\omega_2, \mathbf{k}_2)$ is then mapped on the $\omega = kC_s$ line to identify the decay mode $(\omega_{s2}, \mathbf{k}_{s2})$. The vector relations shown in the figure verify the frequency and wavevector matches in this cascade. A similar procedure starting at the point $(\omega_2, \mathbf{k}_2)$ is applied to determine the plasma modes (ω_3, $\mathbf{k}_3$) and (ω_{s3}, $\mathbf{k}_{s3}$) in the subsequent cascade. As shown, $|k_3|$ is smaller than $|k_2|$ and $|k_2|$ is smaller than $|k_1|$; the available spectral region for cascade diminishes continuously. In other words, not too many cascades can proceed at the same location. However, as the pump wave propagates downward to the lower density region, the parabola of the Langmuir mode in the figure moves down, indicating that additional cascade lines can be accommodated.

In backscatter radar detection, HFPLs have a fixed k value. Therefore, the parabola has to move down slightly anyway in each subsequent cascade to keep the sideband to have a fixed $|k|$. In other words, the cascade lines in the HFPLs, in fact, originate from an extended region below the originating layers of the PDI/OTSI lines in the HFPLs. In the frequency spectrum of HFPLs, the first spectral peak having the highest frequency at $\omega = \omega_1$ is an OTSI line if $\omega_1 = \omega_0$, the heating wave frequency; if ω_1 is downshifted from ω_0 by $\Delta\omega = \omega_0 - \omega_1 = \omega_{s0} = 2k_R C_s$, where k_R is the wavenumber of the probing backscatter radar signal, it is a PDI line; the subsequent spectral peaks at ω_2, ω_3, ... correspond to the first, second, ... cascade lines. Each cascade line is downshifted from the preceding line by $2\omega_{s0}$. For example, if a spectrogram of HFPLs contains 7 spectral peaks starting at $\omega = \omega_0$, then the first two spectral peaks at $\omega = \omega_0$ and $\omega_0 - \omega_{s0}$ are called OTSI and PDI lines, respectively. The remaining 5 spectral peaks at $\omega_0 - (2n + 1)\omega_{s0}, n = 1, \ldots, 5$ are called cascade lines, and their first pump is a sideband of the PDI (the sidebands of the OTSI are slightly off as plasma modes and are usually much weaker). In this case, the spectral width of the HFPLs is about 11 ω_{s0}.

Langmuir cascade is directly of interest in HF heating experiments because it produces HFPLs which are detected by UHF and VHF backscatter radars. However, large amplitude Langmuir/upper hybrid waves can also excite modulation and filamentation instabilities; in essence, both are four-wave interaction processes similar to that of OTSI. In the modulation instability, the decay modes $(\Delta\mathbf{k}, \Delta\omega)$ are long-wavelength low-frequency density fluctuations which modulate the amplitude of the pump wave ($\mathbf{k}$, ω) in space and time, where $|\Delta\mathbf{k}| \ll |\mathbf{k}|$. For example, a Langmuir pump $(\mathbf{k}_{||}, \omega)$ can effectively excite two sidebands ($\mathbf{k}_{||} + \Delta\mathbf{k}_{||}$, $\omega + \Delta\omega$) and ($\mathbf{k}_{||} - \Delta\mathbf{k}_{||}$, $\omega - \Delta\omega$) if $\omega + \Delta\omega \cong [\omega_{pe}^2 + 3(k_{||} + \Delta k_{||})^2 v_{te}^2]^{1/2}$ and $\omega - \Delta\omega \cong [\omega_{pe}^2 + 3(k_{||} - \Delta k_{||})^2 v_{te}^2]^{1/2}$, which can be satisfied under the conditions: $\Delta\omega \cong (\Delta k_{||}/k_{||})(3k_{||}^2 v_{te}^2/\omega)$ and $(\Delta k_{||}/k_{||}) \ll 1$. Thus, the differences between OTSI and modulation instability are summarized to be Re $(\Delta\omega) = 0$ and $(\Delta k_{||}/k_{||}) \gg 1$ in OTSI and Re $(\Delta\omega) \neq 0$ and $(\Delta k_{||}/k_{||}) \ll 1$ in modulation instability, where "Re" represents real part of a complex function. In the filamentation instability, the decay modes $(\Delta\mathbf{k}_{\perp}, i\gamma)$ are purely growing density fluctuations which filament the amplitude of the pump wave $(\mathbf{k}, \omega)$ in space, where $\Delta\mathbf{k}_{\perp} \perp \mathbf{k}$ to keep $|\mathbf{k} + \Delta\mathbf{k}_{\perp}| = |\mathbf{k} - \Delta\mathbf{k}_{\perp}|$.

6.2 Formulation

In the following, parametric excitation of Langmuir/upper hybrid waves, represented by the potential function $\phi(\omega, \mathbf{k})$, and low-frequency plasma waves, represented by the density function $n_s(\omega_s$, $\mathbf{k}_s)$, by EM or Langmuir/upper hybrid pump waves, represented by the electric field function $\mathbf{E}_p(\omega_0$, $\mathbf{k}_p)$, are formulated, where $\mathbf{E}_p, \phi$, and n_s are electric field of a pump wave, ES potential of the Langmuir/upper hybrid sideband, and density perturbation of the low-frequency decay mode, respectively. Langmuir waves can have large oblique propagation angles (with respect to the geomagnetic field $\mathbf{B}_0 = -\hat{\mathbf{z}}\,B_0$), upper hybrid waves have near $90°$ propagation angles, and low-frequency plasma waves include ion acoustic/lower hybrid waves as well as purely growing density striations/FAIs.

6.2.1 *Equations for high-frequency plasma waves*

The coupled mode equation for the Langmuir/upper hybrid sideband is derived from the electron continuity equation (1.20) and momentum equation (1.26), and Poisson equation, which are represented as follows:

$$\frac{\partial\,\delta n_e}{\partial t} + n_0 \boldsymbol{\nabla} \cdot \delta \mathbf{v}_e + \boldsymbol{\nabla} \cdot \langle n_e \mathbf{v}_e \rangle = 0 \tag{6.2}$$

$$\left(\frac{\partial}{\partial t} + \nu_{et}\right)\delta \mathbf{v}_e - |\Omega_e|\delta \mathbf{v}_e \times \hat{\mathbf{z}}$$

$$= -\langle \mathbf{v}_e \cdot \boldsymbol{\nabla}\, \mathbf{v}_e \rangle - 3v_{te}^2 \boldsymbol{\nabla}\frac{\delta n_e}{n_0} - \frac{e}{m_e}\mathbf{E}_h \tag{6.3}$$

and

$$\nabla^2 \phi_h = e\frac{\delta n_e}{\varepsilon_0} \tag{6.4}$$

where $n_e = n_0 + \delta n_e + n_s$; n_0, δn_e, and n_s are the unperturbed plasma density, the electron density perturbation associated with Langmuir/upper hybrid waves, and the low-frequency electron density perturbation, respectively; $v_{te} = (T_e/m_e)^{1/2}$ the electron thermal speed; $\mathbf{E}_h = -\boldsymbol{\nabla}\phi_h$ is the Langmuir/upper hybrid wave field; the adiabatic relationship $\boldsymbol{\nabla}P_e = 3T_e\boldsymbol{\nabla}\delta n_e$ is used; and $\langle\rangle$ stands for a filter, which keeps only terms having the same phase function as that of the wave field $\mathbf{E}_h$; $\nu_{et} = \nu_{en} + \nu_{ei} + \nu_{eL} = \nu_e + \nu_{eL}$ is the effective electron collision frequency, where $\nu_e = \nu_{en} + \nu_{ei}$, ν_{en} is the electron-neutral elastic collision frequency, ν_{ei} the electron–ion Coulomb collision frequency [$\nu_{ei} = 2.632 \times 10^{-6}\,(n_0/T_e^{3/2})$ $\ln\Lambda \cong 39.5 \times 10^{-6}\,(n_0/T_e^{3/2}) \cong 4.87 \times 10^{-7}\,(f_{pe}^2/T_e^{3/2})$, here the Coulomb logarithm $\ln\Lambda \cong 15$ is assumed; n_0 is in m^{-3}, T_e is in K, and f_{pe} is the electron plasma frequency], and ν_{eL} a phenomenological collision frequency to incorporate the electron Landau damping effect

[$\nu_{eL} = (\pi/2)^{1/2}(\omega_{ek}^2\omega_p^2/k_z k^2 v_{te}^3)\exp(-\omega_{ek}^2/2k_z^2 v_{te}^2) = 2\gamma_{ek}$; where γ_{ek} is given in (3.28b) extended to the oblique propagation situation].

6.2.2 *Derivation of the coupled mode equation for the Langmuir/upper hybrid sideband*

With the aid of (6.4), the three orthogonal components of (6.3) are combined into a single scalar equation

$$\begin{aligned}
&\left(\frac{\partial}{\partial t} + \nu_{eh}\right)\left[\left(\frac{\partial}{\partial t} + \nu_{eh}\right)^2 + \Omega_e^2\right]\boldsymbol{\nabla}\cdot\mathbf{v}_e \\
&= \left[\left(\frac{\partial}{\partial t} + \nu_{eh}\right)^2 + \Omega_e^2\right][\omega_p^2 - 3v_{te}^2\nabla^2]\frac{\delta n_e}{n_0} \\
&\quad + \Omega_e^2\left[3v_{te}^2\nabla_\perp^2\frac{\delta n_e}{n_0} - \frac{e}{m_e}\nabla_\perp^2\phi_h\right] \\
&\quad + \left\{|\Omega_e|\left(\frac{\partial}{\partial t} + \nu_{eh}\right)\boldsymbol{\nabla}\cdot\hat{\mathbf{z}}\times[\langle\mathbf{v}_e\cdot\boldsymbol{\nabla}\,\mathbf{v}_e\rangle]\right. \\
&\quad \left. - \left[\left(\frac{\partial}{\partial t} + \nu_{eh}\right)^2\boldsymbol{\nabla} + \Omega_e^2\boldsymbol{\nabla}_z\right]\cdot[\langle\mathbf{v}_e\cdot\boldsymbol{\nabla}\mathbf{v}_e\rangle]\right\} \qquad (6.5)
\end{aligned}$$

where $\omega_p = \omega_{pe} = (n_0 e^2/m_e\varepsilon_0)^{1/2}$, simplifying the notation of the electron plasma frequency.

The RHS of (6.5) is assembled into three groups. The first two groups contain linear response terms and the last one contains coupling terms. The terms in the last group are ascribed to the high-frequency ponderomotive force ($\propto \langle\mathbf{v}_e\cdot\boldsymbol{\nabla}\,\mathbf{v}_e\rangle$), which can be neglected because their contribution to the parametric coupling is much smaller than that from the beating current term (third term) of (6.2). Equations (6.2), (6.4), and (6.5), are then combined to obtain a coupled mode equation of the high-frequency (Langmuir/upper hybrid) sideband $\phi_h(\omega, \mathbf{k})$ to be

$$\left\{\left[\left(\frac{\partial}{\partial t} + \nu_{eh}\right)^2 + \Omega_e^2\right]\left(\frac{\partial^2}{\partial t^2} + \nu_{eh}\partial_t + \omega_p^2 - 3v_{te}^2\nabla^2\right)\nabla^2\right.$$

$$
\begin{aligned}
&- \Omega_e^2(\omega_p^2 - 3v_{te}^2\nabla^2)\nabla_\perp^2\} \phi_h \\
&= -\frac{e}{\varepsilon_0}\left(\frac{\partial}{\partial t} + \nu_{eh}\right)\left[\left(\frac{\partial}{\partial t} + \nu_{eh}\right)^2 + \Omega_e^2\right]\boldsymbol{\nabla}\cdot\langle n_s\mathbf{v}_p\rangle_h \\
&= \omega_p^2\left\{\frac{\omega_h\omega_0}{(\omega_0^2 - \Omega_e^2)}\left[\left(\frac{\partial}{\partial t} + \nu_{eh}\right)^2 + \Omega_e^2\right]\left[\boldsymbol{\nabla} - \frac{\Omega_e^2}{\omega_0^2}\boldsymbol{\nabla}_z\right]\cdot\left\langle\frac{\mathbf{E}_p n_s}{n_0}\right\rangle\right. \\
&\left. + |\Omega_e|\left(\frac{\partial}{\partial t} + \nu_{eh}\right)\hat{\mathbf{z}}\cdot\left\langle\boldsymbol{\nabla}\frac{n_s}{n_0}\times\mathbf{E}_p\right\rangle\right\} \qquad (6.6)
\end{aligned}
$$

where the filter $\langle\rangle_h$ keeps only terms having the same phase function as that of the function ϕ_h on the left-hand side. Although (6.6) is derived from the fluid equations, the kinetic effect of electron Landau damping has been included phenomenologically in the collision frequency.

6.2.3 *One fluid equations*

Both electrons and ions respond effectively to the low-frequency wave fields, hence the formulation for the coupled mode equation of the low-frequency decay mode $n_s(\omega_s, \mathbf{k}_s)$ involve electron and ion fluid equations. Because electrons and ions tend to move together, the formulation is simplified by introducing quasi-neutral condition: $n_{si} \cong n_{se} = n_s$. The ion fluid equations are similar to (6.2) and (6.3), except that the subscript e is changed to i, and the charge $-e$ changed to e. Moreover, the collision terms $\nu_{et}\mathbf{v}_e$ and $\nu_{it}\mathbf{v}_i$ in the electron and ion fluid equations are replaced by $\nu_{ei}(\mathbf{v}_e - \mathbf{v}_i) + (\nu_{en} + \nu_{eL})\mathbf{v}_e$ and $\nu_{ie}(\mathbf{v}_i - \mathbf{v}_e) + \nu_i\mathbf{v}_i$, respectively, where $\nu_i = (\nu_{in} + \nu_{iL})$, ν_{in} is the ion-neutral collision frequency, $\nu_{iL} = 2\gamma_{sk} \cong (\pi/2)^{1/2}(\omega_s^2/k_z v_s)(T_e/T_i)^{3/2}\exp(-\omega_s^2/2k_z^2 v_{ti}^2)$ accounts for the ion Landau damping effect on the low-frequency decay mode, and γ_{sk} is given in (3.30b).

With the aid of $\nu_{ie} = (m_e/m_i)\nu_{ei}$ and $m_e|\Omega_e| = m_i\Omega_i$, and neglecting the electron inertial term and the ion convective term in the momentum equations, these two momentum equations are combined to be

$$\left(\frac{\partial}{\partial t} + \nu_i\right)\mathbf{v}_{is} + \frac{m_e}{m_i}\langle\mathbf{v}_e \cdot \boldsymbol{\nabla}\mathbf{v}_e\rangle - \Omega_i(\mathbf{v}_{es} - \mathbf{v}_{is}) \times \hat{\mathbf{z}} = -C_s^2\boldsymbol{\nabla}\frac{n_s}{n_0}$$

$$(6.7)$$

where $(m_e/m_i)\nu_e \ll \nu_i$ is applied.

6.2.4 Derivation of the coupled mode equation for the low-frequency decay mode

With the aid of $\boldsymbol{\nabla}\cdot\mathbf{v}_{is} = -\partial(n_s/n_0)/\partial t$ from the ion continuity equation, (6.7) becomes

$$\left[\frac{\partial}{\partial t}\left(\frac{\partial}{\partial t} + \nu_i\right) - C_s^2\nabla^2\right]\left(\frac{n_s}{n_0}\right)$$

$$= \frac{m_e}{m_i}\boldsymbol{\nabla}\cdot\langle\mathbf{v}_e \cdot \boldsymbol{\nabla}\mathbf{v}_e\rangle - \Omega_i\boldsymbol{\nabla}\cdot[(\mathbf{v}_{es} - \mathbf{v}_{is}) \times \hat{\mathbf{z}}] \qquad (6.8)$$

where the second term on the RHS of (6.8) can be expressed in terms of (n_s/n_0) from the two equations derived by taking the operations "$\times\hat{\mathbf{z}}$" and "$\boldsymbol{\nabla}\cdot$" on the electron momentum equation (6.3) and ion momentum equation (e.g., (1.21)), respectively.

The coupled mode equation of the low-frequency decay mode (ion acoustic/purely growing/lower hybrid/FAI) is then derived to be

$$\left\{\frac{\partial^3}{\partial t^3}\left\{\left(\frac{\partial}{\partial t} + \nu_{es}\right)\left[\frac{\partial}{\partial t}\left(\frac{\partial}{\partial t} + \nu_i\right) - C_s^2\nabla^2\right] + |\Omega_e|\Omega_i\frac{\partial}{\partial t}\right\}\nabla_\perp^2\right.$$

$$+\Omega_e^2\left\{\left(\frac{\partial^2}{\partial t^2} + \Omega_i^2\right)\left[\frac{\partial}{\partial t}\left(\frac{\partial}{\partial t} + \nu_i\right) - C_s^2\nabla^2\right]\right.$$

$$+\Omega_i^2 C_s^2\nabla_\perp^2\right\}\nabla_z^2\right\}\left(\frac{n_s}{n_0}\right) = \frac{m_e}{m_i}\left[\left(\frac{\partial^2}{\partial t^2} + \Omega_i^2\right)\nabla_z^2 + \frac{\partial^2}{\partial t^2}\nabla_\perp^2\right]$$

$$\times\left\{\left[\frac{\partial}{\partial t}\left(\frac{\partial}{\partial t} + \nu_{es}\right)\boldsymbol{\nabla}_\perp + \Omega_e^2\boldsymbol{\nabla}_z\right]\cdot\mathbf{a}_p + \frac{\partial}{\partial t}\left(\frac{\partial}{\partial t} + \nu_{es}\right)\right.$$

$$\times\nabla_\perp^2\frac{\delta T_e}{m_e} + |\Omega_e|\frac{\partial}{\partial t}\boldsymbol{\nabla}\cdot\left[\mathbf{a}_p \times \hat{\mathbf{z}} - |\Omega_e|\frac{\mathbf{J}_B}{n_0}\right]\right\} \qquad (6.9a)$$

where the coupling terms $\mathbf{a}_p = \langle \mathbf{v}_e \cdot \nabla \mathbf{v}_e \rangle$ and $\mathbf{J}_B = \langle n_e \mathbf{v}_e \rangle$, which have arisen from the plasma nonlinearities, become $\mathbf{a}_p = \langle \mathbf{v}_{e1} \cdot \nabla \mathbf{v}_{e1} \rangle$ and $\mathbf{J}_B = \langle n_{e1} \mathbf{v}_{e1} \rangle$ in parametric excitation, $\mathbf{v}_{e1}$ and n_{e1} are the linear parts of $\mathbf{v}_e$ and n_e, which are the electron velocity and density responses to the total high-frequency wave field $\mathbf{E} = \mathbf{E}_P + \mathbf{E}_h$ and are determined by (6.3) and (6.4), respectively, and $\delta T_e = T_e - T_{e0}$ is the result of the differential Ohmic heating, which is significant only for the field-aligned purely-growing modes and can be evaluated from the electron thermal energy equation (1.27)

$$\frac{\partial T_e}{\partial t} + (2/3)\, T_{e0} \nabla \cdot \mathbf{v}_e$$
$$= \frac{2}{3n_0} \nabla \cdot (\kappa_z \nabla_z + \kappa_\perp \nabla_\perp) T_e - 2\nu_e \frac{m_e}{m_i}(T_e - T_{e0}) + (2/3)\, \nu_e m_e \langle v_e^2 \rangle \tag{6.9b}$$

where $\kappa_z = 3n_0 T_{e0}/2m_e\nu_e$, $\kappa_\perp = (\nu_e/\Omega_e)^2 \kappa_z$, and T_{e0} is the unperturbed electron temperature, and (6.9b) is reduced to be

$$\left[\frac{\partial}{\partial t} - (v_{te}^2 \nu_e/\Omega_e^2)\nabla_\perp^2 + 2\nu_e \frac{m_e}{m_i}\right] \delta T_e \cong (2/3)\, \nu_e m_e \langle v_e^2 \rangle \tag{6.9c}$$

δT_e becomes significant only when the scale lengths of the field-aligned purely-growing modes are large, i.e., $|(v_{te}^2 \nu_e/\Omega_e^2)\nabla_\perp^2| \ll 2\nu_e(m_e/m_i)$. In this scale length regime, $\delta T_e \cong m_i \langle v_e^2 \rangle/3$.

In HF heating experiments, UHF and VHF monostatic (backscatter) radars monitor the excitation of plasma waves. The recorded HFPLs and HFILs are contributed only by those propagating oblique to the geomagnetic field at an angle conjugate to the magnetic dip angle. On the other hand, parametrically excited plasma waves are expected to have broad spectra and large angular distributions. Moreover, upper hybrid waves, lower hybrid waves, and field-aligned density irregularities, which cannot be detected by UHF and VHF backscatter radars, are also expected to be excited in the O-mode HF heating experiments.

6.3 Manley–Rowe Relations

In three-wave parametric decay process, a pump of larger frequency ω_0 transforms into output signals of smaller frequencies ω_1 and ω_s; the power flow P_0 from the pump to the output signals, which have power gains P_1 and P_s, must be balanced. In the lossless system, the power balance requires $P_0 + P_1 + P_s = 0$. Moreover, the changes of the photon/phonon (quasi-particle) numbers among three waves also have to be balanced, namely, $P_1/\omega_1 = P_s/\omega_s = -P_0/\omega_0$. These Manley–Rowe relations show that the power P_s at the decay mode frequency ω_s is smaller than the input power $|P_0|$ by a factor of ω_s/ω_0, while the power P_1 at the idler frequency ω_1 is smaller than $|P_0|$ by a factor of ω_1/ω. In other words, an input signal of larger frequency transforms into output signals of smaller frequencies, and the powers of the output signals get smaller. In the following, the parametric decay of an EM pump into Langmuir wave and ion acoustic wave in unmagnetized lossless plasma is considered to illustrate the Manley–Rowe relations.

6.3.1 *Coupled mode equation for the electromagnetic pump*

With the aid of (1.21), (1.22a) and (1.22b) are combined to be

$$\left(\frac{\partial^2}{\partial t^2} + \omega_p^2 - c^2\nabla^2\right)\mathbf{E}_p = \frac{e}{\varepsilon_0}\frac{\partial}{\partial t}\langle n_s \mathbf{v}_{eh}\rangle_p \tag{6.10}$$

where $\mathbf{v}_{eh}$ is the electron velocity response to the Langmuir wave field $\mathbf{E}_h$, n_s is the density perturbation of the ion acoustic wave, and the filter $\langle\rangle_p$ keeps only terms having the same phase function as that of the pump field $\mathbf{E}_p$ on the left-hand side of (6.10).

6.3.2 *Coupled mode equations for the Langmuir wave and ion acoustic wave*

Set cyclotron frequencies Ω_e and Ω_i and collision frequencies ν_{eh}, ν_{es}, and ν_i equal to zero in (6.6) and (6.9), then the coupled mode equations for the Langmuir wave and ion acoustic wave are obtained respectively to be

$$\left(\frac{\partial^2}{\partial t^2} + \omega_p^2 - 3v_{te}^2\nabla^2\right)\mathbf{E}_h = -\omega_p^2\left(\frac{\omega_1}{\omega_0}\right)\left\langle\frac{\mathbf{E}_p n_s}{n_0}\right\rangle_h \tag{6.11}$$

and

$$\left(\frac{\partial^2}{\partial t^2} + C_s^2\nabla^2\right)\left(\frac{n_s}{n_0}\right) = \frac{m_e}{m_i}\nabla^2\langle v_{eh}v_{ep}\rangle \tag{6.12}$$

6.3.3 *Rate equations for the amplitudes of the three waves*

Substituting $\mathbf{E}_p = \hat{\mathbf{z}}\mathcal{E}_p(t)\exp[i(k_0x - \omega_0 t)] + c.c.$, $\mathbf{E}_h = \hat{\mathbf{z}}\mathcal{E}_h(t)\exp[i(k_1 z - \omega_1 t)] + c.c.$, and $n_s = N_s(t)\exp[i(k_{sx}x + k_{sz}z - \omega_s t)] + c.c.$ into (6.10) to (6.12), with the aid of $v_{ep} = -i(e/m_e\omega_0)\mathcal{E}_p(t)\exp[i(k_0x - \omega_0 t)] + c.c.$ and $v_{eh} = -i(e\omega_1/m_e\omega_p^2)\mathcal{E}_h(t)\exp[i(k_1 z - \omega_1 t)] + c.c.$, where c.c. represents complex conjugate, and neglecting second-order time derivative terms, we obtain

$$\frac{d\mathcal{E}_p(t)}{dt} = -\frac{i}{2}\omega_1\left(\frac{N_s}{n_0}\right)\mathcal{E}_h \tag{6.13}$$

$$\frac{d\mathcal{E}_h(t)}{dt} = -\frac{i}{2}\frac{\omega_p^2}{\omega_0}\left(\frac{N_s^*}{n_0}\right)\mathcal{E}_p \tag{6.14}$$

and

$$\frac{d}{dt}\left(\frac{N_s}{n_0}\right) = -\frac{i}{2}\frac{e^2}{m_e m_i}\left(\frac{k_s^2\omega_1}{\omega_s\omega_0\omega_p^2}\right)\mathcal{E}_h^*\mathcal{E}_p \tag{6.15}$$

In the process of derivation, the dispersion relations $\omega_0 = (\omega_p^2 + k_0^2c^2)^{1/2}$, $\omega_1 = (\omega_p^2 + 3k_1^2v_{te}^2)^{1/2}$, and $\omega_s = k_sC_s$ are employed to finalize the results.

6.3.4 *Energy densities of the waves*

In a dispersive dielectric medium, the electric energy density W_E of the wave (see Sec. 3.5) is given to be

$$W_E = \varepsilon_0 \frac{\partial}{\partial\omega}[\omega\varepsilon(\omega)]\, |\mathcal{E}|^2 \tag{6.16}$$

where the dielectric function $\varepsilon(\omega) = \varepsilon_T(\omega) = 1 - \omega_p^2/\omega_0^2 = k_0^2c^2/\omega_0^2$ for the EM pump; $\varepsilon(\omega) = \varepsilon_h(\omega) = 1 - \omega_p^2/(\omega_1^2 - 3k_1^2v_{te}^2) = 0$ for the Langmuir sideband; and $\varepsilon(\omega) = \varepsilon_s(\omega) = 1 + k_{De}^2/k_s^2 - \omega_{pi}^2/(\omega_s^2 - 3k_s^2v_{ti}^2) = 0$ for the ion acoustic decay mode. The magnetic energy density of the EM wave is given by

$$W_M = \mu_0|\mathcal{H}|^2 = \varepsilon_0\varepsilon_T(\omega)|\mathcal{E}|^2 \tag{6.17}$$

where the relation $\mathcal{H} = \mathcal{E}/\eta$ with the intrinsic impedance $\eta = [\mu_0/\varepsilon_0\varepsilon_T(\omega)]^{1/2}$ is employed.

We now use (6.16) and (6.17) to obtain the energy density W_p of the pump and use (6.16) to obtain the energy densities W_H and W_s of the Langmuir wave and ion acoustic wave. The results are

$$\begin{aligned} W_p &= \varepsilon_0 \frac{\partial}{\partial\omega}\, [\omega\varepsilon_T(\omega)]|\mathcal{E}_p|^2 + \varepsilon_0\varepsilon_T(\omega)|\mathcal{E}_p|^2 \\ &= 2\varepsilon_0|\mathcal{E}_p|^2 = \mathcal{N}_p\hbar\omega_0 \end{aligned} \tag{6.18}$$

$$W_h = \varepsilon_0\frac{\partial}{\partial\omega}[\omega\varepsilon_h(\omega)]|\mathcal{E}_h|^2 = 2\varepsilon_0\frac{\omega_1^2}{\omega_p^2}|\mathcal{E}_h|^2 = \mathcal{N}_h\hbar\omega_1 \tag{6.19}$$

and

$$\begin{aligned} W_s &= \varepsilon_0\frac{\partial}{\partial\omega}\, [\omega\varepsilon_s(\omega)]|\mathcal{E}_s|^2 = 2\varepsilon_0\frac{\omega_s^2}{\omega_{pi}^2}\left(1 + \frac{k_{De}^2}{k_s^2}\right)^2 |\mathcal{E}_s|^2 \\ &= 2m_in_0\frac{\omega_s^2}{k_s^2}\left|\frac{N_s}{n_0}\right|^2 = \mathcal{N}_s\hbar\omega_s \end{aligned} \tag{6.20}$$

where the relation $ik_s\mathcal{E}_s = e(N_{si} - N_{se})/\varepsilon_0 = eN_{si}/\varepsilon_0(1 + k_{De}^2/k_s^2)$ obtained from the Poisson equation and $N_{si} = N_s$, a simplification of the notation, are applied in (6.20); $\mathcal{N}_p$, $\mathcal{N}_h$, and $\mathcal{N}_s$ are the photon and phonon densities of the pump wave, Langmuir wave, and ion acoustic wave; and $\hbar = h/2\pi = 6.582 \times 10^{-16}$ eV·s is the reduced Planck constant.

6.3.5 *Demonstration of Manley–Rowe relations*

With the aid of (6.13) to (6.15) and (6.18) to (6.20), we have

$$\begin{aligned} P_0 &= \frac{dW_p}{dt} = 2\varepsilon_0 \mathcal{E}_p \frac{d\mathcal{E}_p^*}{dt} + c.c. \\ &= i\,\varepsilon_0\,\omega_1\,\mathcal{E}_p \left(\frac{N_s^*}{n_0}\right) \mathcal{E}_h^* + c.c. \end{aligned} \tag{6.21}$$

$$\begin{aligned} P_1 &= \frac{dW_h}{dt} = 2\varepsilon_0 \frac{\omega_1^2}{\omega_p^2} \mathcal{E}_h \frac{d\mathcal{E}_h^*}{dt} + c.c. \\ &= -i\varepsilon_0 \frac{\omega_1^2}{\omega_0} \mathcal{E}_p \left(\frac{N_s^*}{n_0}\right) \mathcal{E}_h^* + c.c. \end{aligned} \tag{6.22}$$

and

$$\begin{aligned} P_s &= \frac{d}{dt} W_s = 2m_i \frac{\omega_s^2}{k_s^2} N_s^* \frac{d}{dt}\left(\frac{N_s}{n_0}\right) + c.c. \\ &= -i\varepsilon_0 \frac{\omega_1 \omega_s}{\omega_0}\, \mathcal{E}_p \left(\frac{N_s^*}{n_0}\right) \mathcal{E}_h^* + c.c. \end{aligned} \tag{6.23}$$

It shows that $P_0 + P_1 + P_s = (\omega_0 - \omega_1 - \omega_s)[i\varepsilon_0(\omega_1/\omega_0)\mathcal{E}_p(N_s^*/n_0)$ $\mathcal{E}_h^* + c.c.] = 0$, a balance of power flow; and $P_1/\omega_1 = -i\varepsilon_0(\omega_1/\omega_0)\, \mathcal{E}_p$ $(N_s^*/n_0)\, \mathcal{E}_h^* + c.c. = P_s/\omega_s = -P_0/\omega_0$, which are also expressed to be $d/dt\ (\mathcal{N}_p + \mathcal{N}_h) = 0 = d/dt(\mathcal{N}_p + \mathcal{N}_s) = d/dt(\mathcal{N}_h - \mathcal{N}_s)$, conservation of quasi-particle number.

6.3.6 *Analysis*

Let $\mathcal{E}_p = |\mathcal{E}_p|\exp(i\theta_p)$, $N_s/n_0 = |N_s/n_0|\exp(i\theta_s)$, and $\mathcal{E}_h = |\mathcal{E}_h|\exp(i\theta_h)$, which, with the aid of (6.18) to (6.20), become $\mathcal{E}_p = (\hbar\omega_0/2\varepsilon_0)^{1/2}\mathcal{N}_p^{1/2}$ $\exp(i\theta_p)$, $N_s/n_0 = (k_s/\omega_s)(\hbar\omega_s/2m_i n_0)^{1/2}\mathcal{N}_s^{1/2}\exp(i\theta_s)$, and $\mathcal{E}_h = (\omega_p/$

$\omega_1)(\hbar\omega_1/2\varepsilon_0)^{1/2}\mathcal{N}_h^{1/2}\exp(i\theta_h)$. Thus, (6.21) to (6.23) become

$$\frac{d\,\mathcal{N}_p}{dt} = -A\mathcal{N}_p^{1/2}\mathcal{N}_h^{1/2}\mathcal{N}_s^{1/2} \tag{6.24}$$

$$\frac{d\,\mathcal{N}_h}{dt} = A\mathcal{N}_p^{1/2}\mathcal{N}_h^{1/2}\mathcal{N}_s^{1/2} \tag{6.25}$$

and

$$\frac{d\,\mathcal{N}_s}{dt} = A\mathcal{N}_p^{1/2}\mathcal{N}_h^{1/2}\mathcal{N}_s^{1/2} \tag{6.26}$$

where $A = k_s\omega_p(\hbar\omega_1/2n_0m_i\omega_0\omega_s)^{1/2}\sin(\theta_p - \theta_h - \theta_s)$.

We now use the relations:

$$\mathcal{N}_p + \mathcal{N}_h = \mathcal{N}_{p0} + \mathcal{N}_{h0} = \mathcal{N}_{10}$$

and

$$\mathcal{N}_p + \mathcal{N}_s = \mathcal{N}_{p0} + \mathcal{N}_{s0} = \mathcal{N}_{20},$$

where $\mathcal{N}_{p0}$, $\mathcal{N}_{h0}$, and $\mathcal{N}_{s0}$ are the initial values and $\mathcal{N}_{10}$ and $\mathcal{N}_{20}$ are constants.

Equation (6.24) is converted to be

$$\frac{dy}{dt} = -H(1 - y^2)^{1/2}(1 - \gamma^2y^2)^{1/2} \tag{6.27}$$

where $y = (\mathcal{N}_p/\mathcal{N}_{10})^{1/2}$, $H = (A/2)\mathcal{N}_{20}^{1/2}$, and $\gamma = (\mathcal{N}_{10}/\mathcal{N}_{20})^{1/2}$.

The solution of (6.27) is a Jacobi elliptic function $sn(H(t_0-t),\gamma)$, i.e.,

$$y(t) = sn(H(t_0 - t), \gamma),$$

where $sn(Ht_0, \gamma) = (\mathcal{N}_{p0}/\mathcal{N}_{10})^{1/2}$, i.e., $t_0 = (1/H)sn^{-1}(\mathcal{N}_{p0}/\mathcal{N}_{10})^{1/2}$.

Thus,

$$\mathcal{N}_p = \mathcal{N}_{10}\, sn^2(H(t_0 - t), \gamma),$$

$$\mathcal{N}_h = \mathcal{N}_{10}\, cn^2(H(t_0 - t), \gamma),$$

and

$$\mathcal{N}_{\rm s} = (\mathcal{N}_{\rm s0} - \mathcal{N}_{\rm h0}) + \mathcal{N}_{10}\, cn^2(H(\rm t_0 - t), \gamma) = \mathcal{N}_{20}\, dn^2(H(\rm t_0 - t), \gamma),$$

where $cn^2(H(\rm t_0-t),\gamma) = 1 - sn^2(H(\rm t_0-t),\gamma)$ and $dn^2(H(\rm t_0-t),\gamma) = 1 - \gamma^2 sn^2(H(\rm t_0 - t),\gamma)$.

Problems

P6.1. Apply (6.6) & (6.9) to derive the couple mode equations for parametric excitation in unmagnetized plasma.

P6.2. Use the low-frequency ion mode equation derived in Problem P6.1 to show that high-frequency Langmuir waves induce density striation (i.e., no real frequency) given to be

$$\frac{\rm n_s}{\rm n_0} = -\frac{\rm m_e}{\rm m_i}\left(\frac{1}{\rm C_s^2}\right)\left\langle\frac{\rm v_{eh}^2}{2}\right\rangle$$

where $\rm v_{eh}$ is the electron velocity driven by the Langmuir wave field $\mathbf{E}_{\rm h}$.
[Hint: $\langle \mathbf{v}_{\rm e} \cdot \boldsymbol{\nabla}\mathbf{v}_{\rm e}\rangle = \langle \boldsymbol{\nabla}\rm v_e^2/2\rangle$ in unmagnetized plasma]

P6.3. Consider collisionless and unmagnetized plasma; show that

(1) Eqs. (6.2) to (6.4) can be combined to be

$$\left(\frac{\partial^2}{\partial \rm t^2} + \omega_{\rm p}^2 - 3\rm v_{te}^2\nabla^2\right)\mathbf{E}_{\rm h} = \frac{\rm e}{\varepsilon_0}\left[\frac{\partial}{\partial \rm t}\langle \rm n_e \mathbf{v}_e\rangle_h - n_0\boldsymbol{\nabla}\left\langle\frac{v_e^2}{2}\right\rangle_h\right] \qquad \text{(P6.1)}$$

(2) $|\rm n_0\boldsymbol{\nabla}\langle v_e^2/2\rangle_h| \ll |\partial_t\langle n_e\mathbf{v}_e\rangle_h|$
[Hint: $\langle \rm n_e\mathbf{v}_e\rangle_h = \langle n_s\mathbf{v}_{ep}\rangle_h$ and $\langle \rm v_e^2/2\rangle_h = \langle v_{es}v_{ep}\rangle_h$]

(3) Under the condition (2), (P6.1) reduces to (6.11).

P6.4. Show the assumptions used to derive (6.13) to (6.15) from (6.10) to (6.12).

P6.5. Apply mode coupling relation (4.36) to formulate parametric coupling. Consider PDI/OTSI in a collisionless and unmagnetized plasma, where the fields of the dipole pump $\mathcal{E}_{\rm p}(0,\omega_0)$, Langmuir sidebands $\mathcal{E}_\ell^{\pm}(\mathbf{k}_\pm,\omega_\pm)$, and decay mode $\mathcal{E}_{\rm s}(\mathbf{k}_{\rm s},\omega_{\rm s})$ are all parallel to each other. The frequency and wavevector matching conditions are $\omega_\pm = \omega_0 \pm \omega_s^\pm$ and $\mathbf{k}_\pm = \pm\mathbf{k}_s$, where $\omega_s^+ = \omega_s$ and $\omega_s^- = \omega_s^*$. Show that the dispersion equation is given to be

$$\varepsilon_s(\mathbf{k}_s, \omega_s) = \left(\frac{e}{m_e}\right)^2 \frac{k_{de}^4}{k_s^2\omega_0^4}\left[\frac{1}{\varepsilon_L(\mathbf{k}_+, \omega_+)} + \frac{1}{\varepsilon_L^*(\mathbf{k}_-, \omega_-)}\right]|\mathcal{E}_p|^2$$

where $\varepsilon_s(\mathbf{k}_s, \omega_s)$, and $\varepsilon_L(\mathbf{k}_\pm, \omega_\pm)$ are the ES plasma dielectric functions in the low-frequency and high-frequency regimes. [Hint: set $\mathcal{E}_2^\pm(k_\pm, \omega_\pm) = \mathcal{E}_\ell^\pm(\mathbf{k}_\pm, \omega_\pm)$, $\mathcal{E}_{11} = \mathcal{E}_p$ and $\mathcal{E}_{12} = \mathcal{E}_s$ in (4.36) as well as set $\mathcal{E}_2^\pm(k_\pm, \omega_\pm) = \mathcal{E}_s(\mathbf{k}_s, \omega_s), \mathcal{E}_{11}\mathcal{E}_{12}^+ = \mathcal{E}_p^*\mathcal{E}_\ell^+(\mathbf{k}_+, \omega_+)$, and $\mathcal{E}_{11}\mathcal{E}_{12}^- = \mathcal{E}_p\mathcal{E}_\ell^{-*}(\mathbf{k}_-, \omega_-)$ in (4.36), then combine two equations to obtain a dispersion equation]

P6.6. Consider a special case that $\mathcal{E}_p = i|\mathcal{E}_p|$, $N_s/n_0 = |N_s/n_0|$, and $\mathcal{E}_h = |\mathcal{E}_h|$, in the set of coupled equations (6.21) to (6.23).

(1) Show that Eqs. (6.21) and (6.23) can be expressed to be

$$\frac{d\,\mathcal{N}_p}{dt} = -A_1\mathcal{N}_p^{1/2}\mathcal{N}_h^{1/2}\mathcal{N}_s^{1/2}$$

$$\frac{d\,\mathcal{N}_h}{dt} = A_1\mathcal{N}_p^{1/2}\mathcal{N}_h^{1/2}\mathcal{N}_s^{1/2}$$

and

$$\frac{d\,\mathcal{N}_s}{dt} = A_1\mathcal{N}_p^{1/2}\mathcal{N}_h^{1/2}\mathcal{N}_s^{1/2}$$

where $A_1 = k_s\omega_p(\hbar\omega_1/2n_0m_i\omega_0\omega_s)^{1/2}$.

(2) In the special case that $\mathcal{N}_p(0) = \mathcal{N}_{p0}, \mathcal{N}_h(0) = \mathcal{N}_{h0} = \mathcal{N}_0$, and $\mathcal{N}_s(0) = \mathcal{N}_{s0} = \mathcal{N}_0$, i.e., $\mathcal{N}_{h0} = \mathcal{N}_{s0} = \mathcal{N}_0$, show that the set of equations can be consolidated to a single equation

$$\frac{d\,y_1}{dt} = -H_1(1 - y_1^2)$$

where $y_1 = [\mathcal{N}_p/(\mathcal{N}_{p0} + \mathcal{N}_0)]^{1/2}$.

Chapter 7

Parametric Instabilities Excited in High-Frequency Heating Experiments

Decays of an O-mode EM dipole pump $\mathbf{E}_\mathrm{p}(\omega_0, \mathbf{k}_\mathrm{p}=0)$ into Langmuir/upper hybrid sidebands $\phi_\pm(\omega_\pm, \mathbf{k}_\pm)$ and purely growing/FAI decay mode or ion acoustic/lower hybrid decay mode $\mathrm{n_s}(\omega_\mathrm{s}, \mathbf{k}_\mathrm{s})$ in the spatial region below the high frequency (HF) reflection height/upper hybrid resonance layer are studied; $\phi_\pm$ and $\mathrm{n_s}$ denote the sidebands' electrostatic potentials and decay mode's density perturbation, respectively, and ϕ_- is neglected in the three wave coupling (e.g., Parametric decay instability (PDI). Applying the same notations in the two cases of (6.1a) and (6.1b), the frequency and wavevector matching conditions (6.1) become

$$\omega_\pm = \omega = \omega_0 - \omega_\mathrm{s}^* \quad \text{and} \quad \mathbf{k}_+ = \mathbf{k} = -\mathbf{k}_- = -\mathbf{k}_\mathrm{s},$$

and the potential functions

$$\phi_+ = \phi(\omega, \mathbf{k}_1) \quad \text{and} \quad \phi_- = \phi'(\omega, \mathbf{k}_1' = -\mathbf{k}_1);$$

where $\mathbf{k} = \hat{\mathbf{z}}\,\mathrm{k_z} + \hat{\mathbf{x}}\,\mathrm{k}_\perp$.

In this spatial region the O-mode heater field is given to be

$$\mathbf{E}_\mathrm{p} = (\hat{\mathbf{x}} + \mathrm{i}\hat{\mathbf{y}})\mathrm{E}_{\mathrm{p}\perp} + \hat{\mathbf{z}}\,\mathrm{E}_\mathrm{pz} + \mathrm{c.c.} \tag{7.1}$$

where $\mathrm{E}_{\mathrm{p}\perp,\mathrm{z}} = E_{\mathrm{p}\perp,\mathrm{z}}\exp(-\mathrm{i}\omega_0\mathrm{t})$, $E_{\mathrm{p}\perp,\mathrm{z}} = \mathrm{E}_{0\perp,\mathrm{z}}/2$, and c.c. represents complex conjugate.

It is noted that $E_{0\perp,z}$ varies with location (i.e., altitude); for instance, near the HF reflection height $\mathbf{E}_p \cong \hat{\mathbf{z}} E_{pz} + \text{c.c.} \cong \hat{\mathbf{z}} E_{0z} \cos\omega_0 t$ and $E_{0\perp} \cong 0$; and in the upper hybrid resonance region $\mathbf{E}_p \cong (\hat{\mathbf{x}} + i\hat{\mathbf{y}})E_{p\perp} + \text{c.c.} \cong E_{0\perp}(\hat{\mathbf{x}}\cos\omega_0 t + \hat{\mathbf{y}}\sin\omega_0 t)$ and $E_{0z} \cong 0$.

With the aid of (7.1), (6.6) and (6.9) become

$$\left\{\left[\left(\frac{\partial}{\partial t} + \nu_{eh}\right)^2 + \Omega_e^2\right]\left(\frac{\partial^2}{\partial t^2} + \nu_{eh}\frac{\partial}{\partial t} + \omega_p^2 - 3v_{te}^2\nabla^2\right)\nabla_z^2 \right.$$
$$+ \left(\frac{\partial}{\partial t} + \nu_{eh}\right)\left[\left(\frac{\partial}{\partial t} + \nu_{eh}\right)\left(\frac{\partial^2}{\partial t^2} + \nu_{eh}\frac{\partial}{\partial t} + \omega_p^2 - 3v_{te}^2\nabla^2\right)\right.$$
$$\left.\left. + \Omega_e^2\frac{\partial}{\partial t}\right]\nabla_\perp^2\right\}\phi_\pm = \omega_p^2\left\{\left[\left(\frac{\partial}{\partial t} + \nu_{eh}\right)^2 + \Omega_e^2\right]\frac{\partial}{\partial z}\left(\frac{E_{pz}n_{s\pm}}{n_0}\right)\right.$$
$$\left. + \left(\frac{\partial}{\partial t} + \nu_{eh}\right)\left(\frac{\partial}{\partial x} + i\frac{\partial}{\partial y}\right)\left(\frac{\partial}{\partial t} + \nu_{eh} + i|\Omega_e|\right)\left(\frac{E_{p\perp}n_{s\pm}}{n_0}\right)\right\} \tag{7.2}$$

and

$$\left\{\left[\frac{\partial}{\partial t}\left(\frac{\partial}{\partial t} + \nu_i\right) - C_s^2\nabla^2\right]\left[\frac{\partial}{\partial t}\left(\frac{\partial}{\partial t} + \nu_{es}\right)\nabla_\perp^2 + \Omega_e^2\nabla_z^2\right]\right.$$
$$\left. + |\Omega_e|\Omega_i\frac{\partial^2}{\partial t^2}\nabla_\perp^2\right\}\left(\frac{n_s}{n_0}\right) = -\frac{m_e}{m_i}\left(\frac{e}{m_e}\right)^2\left\{\frac{1}{(\omega^2 - \Omega_e^2)}\frac{\partial}{\partial t}\right.$$
$$\left. \times\left[\frac{\partial}{\partial t} + \nu_{es} + i\frac{\Omega_e^2(\omega_u^2 - \omega^2)}{\omega_p^2\omega}\right]\nabla_\perp^2 + \frac{\Omega_e^2}{\omega^2}\nabla_z^2\right\}$$
$$\times\left\{\left[\frac{\omega}{(\omega + |\Omega_e|)}\left(\frac{\partial}{\partial x} + i\frac{\partial}{\partial y}\right)E_{p\perp} + \frac{\partial E_{pz}}{\partial z}\right]\nabla^2\phi_+^*\right.$$
$$\left. + \left[\frac{\omega}{(\omega + |\Omega_e|)}\left(\frac{\partial}{\partial x} - i\frac{\partial}{\partial y}\right)E_{p\perp}^* + \frac{\partial E_{pz}^*}{\partial z}\right]\nabla^2\phi_-\right\}$$
$$+ \frac{m_e}{m_i}\frac{\partial}{\partial t}\left(\frac{\partial}{\partial t} + \nu_{es}\right)\nabla_\perp^4\left(\frac{\delta T_e}{m_e}\right) \tag{7.3}$$

where $n_{s+}^* = n_s(\omega_s, \mathbf{k}_s = -\mathbf{k}_1) = n_{s-}$, and $[\omega_h\omega_0/(\omega_0^2 - \Omega_e^2)]\ [(\partial/\partial t + \nu_{eh})^2 + \Omega_e^2] \cong (\partial/\partial t + \nu_{eh})^2$ is assumed.

The last term on the RHS of (7.3) is included only when n_s is field-aligned density striation. When $|k^2 v_{te}^2 \nu_e/\Omega_e^2| \ll 2\nu_e(m_e/m_i)$, $\delta T_e \cong m_i\langle v_e^2\rangle/3 = -(e/m_e)^2(\omega_0 + |\Omega_e|)^{-2}(2m_i/3)(E_{p\perp}\nabla_\perp\phi_+^* + E_{p\perp}^*\nabla_\perp\phi_-)$. In the following analysis, this term will be neglected.

Equations (7.2) and (7.3) are analyzed in the k–ω domain, where the spatial and temporal variation of physical functions in (7.2) and (7.3) are set to have the form of $p = p\exp[i(\boldsymbol{\kappa}\cdot\mathbf{r} - \varpi t)]$, where $\boldsymbol{\kappa}$ and $\varpi = \varpi_r + ir$ are the appropriate wavevector and complex frequency of each physical quantity. Thus, (7.2) and (7.3) are converted to the coupled algebraic equations to be

$$[\omega^2 + i\nu_{eh}\omega - (\omega_k^2 + \Omega_e^2\sin^2\vartheta)]\,k^2\phi_\pm$$
$$= \pm i\omega_p^2\left[\frac{(\omega^2 - \Omega_e^2)}{(\omega^2 - \Omega_e^2\cos^2\theta)}\right]$$
$$\times\left\{(k_z E_{pz}) + \frac{\omega}{(\omega + |\Omega_e|)}[(k_x + ik_y)E_{p\perp}]\right\}\left(\frac{n_{s\pm}}{n_0}\right) \quad (7.4)$$

and

$$\left\{\omega_s(\omega_s + i\nu_i)\left[\omega_s\left(\omega_s + i\nu_{es} + i\frac{\nu_i}{\xi}\right) - (k^2C_s^2 + |\Omega_e|\Omega_i\xi)\right]\sin^2\theta\right.$$
$$\left. + \Omega_e^2 k^2 C_s^2\cos^2\theta\right\}\left(\frac{n_s}{n_0}\right) = -i\frac{m_e}{m_i}\Omega_e^2\left(\frac{ke}{m_e\omega}\right)^2$$
$$\times\left\{\left[\frac{\omega\omega_s(\omega_u^2 - \omega^2)}{\omega_p^2(\omega^2 - \Omega_e^2)}\right]\sin^2\theta + \cos^2\theta\right\}$$
$$\times\left\{k_z(E_{pz}\phi_+^* + E_{pz}^*\phi_-) + \frac{\omega}{(\omega + |\Omega_e|)}[k_x(E_{p\perp}\phi_+^* + E_{p\perp}^*\phi_-]\right.$$
$$\left. + ik_y(E_{p\perp}\phi_+^* - E_{p\perp}^*\phi_-)]\right\} \quad (7.5)$$

where $\sin^2\vartheta = k_\perp^2/k^2, \omega_k^2 = \omega_{ek}^2 = \omega_p^2 + 3k^2v_{te}^2$, and $\omega_u^2 = \omega_{uH}^2 = \omega_p^2 + \Omega_e^2$; (7.4) is simplified with the condition $(\Omega_e/\omega)^4 \ll 1$; $\xi = 1 + \zeta = 1 + (m_i/m_e)\cos^2\vartheta$.

In the absence of the pump field (HF heater wave), (7.4) and (7.5) reduce to the dispersion equations of the HF and low-frequency plasma modes engaged in the considered parametric couplings. The eigen-frequencies ω_r and ω_{sr} of the high-frequency and low-frequency plasma modes are derived to be

$$\omega_r \cong \begin{cases} \omega_{k\theta} = (\omega_p^2 + 3k^2 v_{te}^2 + \Omega_e^2 \sin^2 \vartheta)^{1/2} & \text{for } \left(\dfrac{k_z}{k_\perp}\right)^2 \gg \dfrac{m_e}{m_i} \\ \omega_{uk} = (\omega_p^2 + \Omega_e^2 + 3k^2 v_{te}^2)^{1/2} & \text{for } \left(\dfrac{k_z}{k_\perp}\right)^2 \ll 1 \end{cases} \tag{7.6}$$

and

$$\omega_{sr} \cong \begin{cases} kC_s & \text{for } \left(\dfrac{k_z}{k_\perp}\right)^2 \gg \dfrac{m_e}{m_i} \\ \omega_{LK} & \text{for } \left(\dfrac{k_z}{k_\perp}\right)^2 \ll 1 \end{cases} \tag{7.7}$$

where $\omega_{Lk} = (|\Omega_e|\Omega_i\xi + k^2C_s^2/\xi)^{1/2}$; ω_r is the eigen-frequency of the electron plasma mode (Langmuir wave) for $(k_z/k_\perp)^2 \gg m_e/m_i$, and of the upper hybrid mode for $(k_z/k_\perp)^2 \ll 1$. In the regions near the HF reflection layer and near the upper hybrid resonance layer, the HF heater wave parametrically excites Langmuir waves and purely growing density striation, and upper hybrid waves and field-aligned density irregularities, respectively. In the region between the HF reflection layer and upper hybrid resonance layer, the HF heater wave excites Langmuir waves and ion acoustic waves parametrically and excites upper hybrid waves and lower hybrid waves parametrically in the region below the upper hybrid resonance layer.

In the following, the coupled mode algebraic equations (7.4) and (7.5) are employed to analyze parametric instabilities relevant to the experimental observations presented in Chapter 5.

7.1 Oscillating Two-Stream Instability and Parametric Decay Instability Near the High-Frequency Reflection Height

As the RH circularly polarized HF heater wave propagates to the region near the reflection height, it converts to the O-mode with electric field $\mathbf{E}_p \cong \hat{\mathbf{z}}\, E_{0z} \cos \omega_0 t$.

7.1.1 *Oscillating two-stream instability — excitation of Langmuir waves together with purely growing density striations by the high-frequency heater wave*

This process involves two Langmuir sidebands $\phi_\pm(\omega, \pm\mathbf{k})$; the coupled mode equations are deduced from (7.4) to be

$$[\omega^2 + i\nu_{eh}\omega - (\omega_k^2 + \Omega_e^2 \sin^2 \vartheta)]k^2\phi_\pm$$
$$= \pm i\omega_p^2 \left[\frac{(\omega^2 - \Omega_e^2)}{(\omega^2 - \Omega_e^2 \cos^2 \vartheta)}\right] \left(\frac{k_z E_{0z}}{2}\right) \left(\frac{n_{s\pm}}{n_0}\right) \quad (7.8)$$

The parallel (to the magnetic field) component k_{sz} of the wavevector of the short-scale purely growing density striation ($\omega_s = i\gamma_s, \mathbf{k}_s$) is not negligibly small, and thus the short scale purely growing density striation is mainly driven by the parallel component of the ponderomotive force induced by the HF wave fields. Using the relation $|\Omega_e|\Omega_i\xi \sim \Omega_e^2 k_z^2/k_\perp^2 \gg k^2C_s^2$, the coupled mode equation for the purely growing density striation is deduced from (7.5) to be

$$[\omega_s(\omega_s + i\nu_i) - k^2C_s^2]\left(\frac{n_s}{n_0}\right)$$
$$= \frac{i}{2}\left(\frac{e^2 k_z k^2}{m_e m_i \omega_0^2}\right)(E_{0z}\phi_+^* + E_{0z}^*\phi_-) \quad (7.9)$$

Set $\omega_s = i\gamma_s$ and $\omega = \omega_0 + i\gamma_s$, (7.8) and (7.9) are combined to be

$$\begin{aligned}\{(\gamma_s^2 + \Omega_i^2)[\gamma_s(\gamma_s + \nu_i) + k^2C_s^2] - \Omega_i^2 k_\perp^2 C_s^2\}\\ = \frac{e^2}{2m_e m_i} k^2 \cos^2\theta \frac{\Delta\omega^2(\gamma_s^2 + \Omega_i^2\cos^2\vartheta)}{[\Delta\omega^4 + \omega_0^2(2\gamma_s + \nu_{eh})^2]} |E_{0z}|^2 \qquad (7.10)\end{aligned}$$

where $\Delta\omega^2 = \omega_p^2 + 3k^2v_{te}^2 + \Omega_e^2 \sin^2\vartheta - \omega_0^2$.

Set $\gamma_s = 0$ in (7.10), then the threshold field is obtained to be

$$|E_{0z}(\vartheta)|_{th} = \left(\frac{2m_e m_i}{e^2}\right)^{1/2} \left[\frac{(\Delta\omega^4 + \omega_0^2\nu_{eh}^2)}{\Delta\omega^2\cos^2\vartheta}\right]^{1/2} C_s \qquad (7.11)$$

Equation (7.11) shows that the threshold field of Oscillating two-stream instability (OTSI) varies with the propagation angle θ and wavelength λ of the Langmuir sidebands as well as the location of excitation (i.e., $\Delta\omega^2$). There is a preferential height layer at altitude $h(k, \vartheta)$ to excite (k, ϑ) lines, where $\omega_p^2(h) = \omega_p^2(k, \vartheta) = \omega_0(\omega_0 + \nu_e) - 3k^2v_{te}^2 - \Omega_e^2\sin^2\vartheta$, $\Delta\omega^2(k, \vartheta) = \omega_0\nu_{eh}$, and the threshold field (7.11) is the minimum

$$|E_{otsi}(k, \vartheta)|_m = 2\left(\frac{m_e m_i}{e^2}\right)^{1/2} \frac{(\omega_0\nu_{eh})^{1/2}}{\cos\vartheta} C_s \qquad (7.12)$$

7.1.2 *Parametric decay instability — decay of high-frequency heater to Langmuir sideband $\phi(\omega, \mathbf{k})$ and ion acoustic wave $n_s(\omega_s, \mathbf{k}_s)$*

The coupled mode equations (7.4) and (7.5) reduce to

$$\begin{aligned}[\omega(\omega + i\nu_{eh}) - \omega_{k\vartheta}^2]\phi\\ = i\left[\frac{(\omega_0^2 - \Omega_e^2)}{(\omega_0^2 - \Omega_e^2\cos^2\vartheta)}\right]\left(\frac{k_z}{2k^2}\right)\omega_p^2 E_{0z}\left(\frac{n_s^*}{n_0}\right) \qquad (7.13)\end{aligned}$$

and

$$[\omega_s(\omega_s + i\nu_i) - k^2C_s^2]\left(\frac{n_s}{n_0}\right) = i\frac{e^2}{2m_e m_i}\left(\frac{k_z k^2}{\omega^2}\right) E_{0z}\phi^* \qquad (7.14)$$

Equations (7.13) and (7.14) are combined to obtain the dispersion relation

$$[\omega(\omega + i\nu_{eh}) - \omega_{k\vartheta}^2][\omega_s^*(\omega_s^* - i\nu_i) - k^2C_s^2]$$
$$= \frac{e^2}{4m_e m_i}\left[\frac{(\omega_0^2 - \Omega_e^2)}{(\omega_0^2 - \Omega_e^2\cos^2\vartheta)}\right]\left(\frac{k_z^2\omega_p^2}{\omega_0^2}\right)|E_{0z}|^2 \qquad (7.15)$$

We now set $\omega = \omega_r$ and $\omega_s = \omega_{sr}$ in (7.15), i.e., the growth rate $\gamma_k = 0$, to evaluate the threshold field $E_{0zth}(k, \vartheta)$ of the instability excited at height h_1, where $\omega_p^2(h_1) = \omega_r^2 - 3k_1^2 v_{te}^2 - \Omega_e^2 \sin^2\vartheta_1$ is the matching height of a specific Langmuir line (k_1, ϑ_1). In this case, the sideband and decay wave of the instability are driven waves, rather than eigen-modes of the plasma. The threshold field of the instability is obtained to be

$$|E_{0z}(k, \vartheta; k_1, \vartheta_1)|_{th}$$
$$= 2\left(\frac{m_e m_i}{e^2}\right)^{1/2}\left[\frac{(\omega_0^2 - \Omega_e^2\cos^2\vartheta)}{(\omega_0^2 - \Omega_e^2)}\right]^{1/2}$$
$$\times\left(1 + \frac{\Delta\omega_1^4}{\omega_0^2\nu_{eh}^2}\right)^{1/2}\frac{(\nu_{eh}\nu_i\omega_{sr}\omega_0^3)^{1/2}}{k\cos\vartheta\omega_p} \qquad (7.16)$$

where $\Delta\omega_1^2 = \omega_{k\theta}^2 - \omega_r^2 = 3(k^2 - k_1^2)v_{te}^2 + \Omega_e^2(\sin^2\vartheta - \sin^2\vartheta_1)$; $\omega_{sr}^2 = k^2C_s^2 - \omega_{sr}\nu_i\Delta\omega_1^2/\omega_r\nu_{eh}$.

It is shown in (7.16) that the threshold field varies with the propagation angle ϑ and wavelength λ of the Langmuir sideband as well as the location of excitation. When the instability is excited at the matching height h of its Langmuir sideband (k, θ), i.e., $\omega_r = \omega_{k\theta}$ and thus $\Delta\omega_1 = 0$, it has the minimum threshold

$$|E_{pdi}(k, \vartheta)|_m$$
$$= 2\left(\frac{m_e m_i}{e^2}\right)^{1/2}\left[\frac{(\omega_0^2 - \Omega_e^2\cos^2\vartheta)}{(\omega_0^2 - \Omega_e^2)}\right]^{1/2}\frac{(\nu_{eh}\nu_i\omega_{sr}\omega_0^3)^{1/2}}{k\cos\vartheta\omega_p} \qquad (7.17)$$

Equations (7.11) and (7.16) show that as the oblique propagation angle ϑ of OTSI- and PDI-excited plasma waves increases, their preferential excitation layers move downward and the threshold fields (7.12) and (7.17) increase. When the heater wave field E_0 is large, the spectral lines of the Langmuir sidebands excited by OTSI and PDI are expected to establish angular (ϑ) and spectral (k) distributions, as well as spatial (h) distributions in an extended region. The altitude region of the OTSI is higher and narrower than that of the PDI.

7.2 Upper Hybrid Oscillating Two-Stream Instability and Parametric Decay Instability Excited Below the Upper Hybrid Resonance Layer

7.2.1 *Oscillating two-stream instability — excitation of upper hybrid waves together with field-aligned density irregularities by the high-frequency heater*

The upper hybrid resonance region is located below the O-mode HF reflection height and is accessible by the O-mode HF heater wave. At high latitude, such as at Tromso, Norway, and Gakona, Alaska, RH circularly polarized heater can be transmitted along the geomagnetic field. The wave fields in the region below the upper hybrid resonance region still remain to be RH circular polarization, i.e., $\mathbf{E}_p = (\hat{\mathbf{x}} + i\hat{\mathbf{y}})(E_{0\perp}/2)\exp(-i\omega_0 t) + \text{c.c.}$, which decays into two upper hybrid sidebands $\phi_\pm(\omega_\pm, \mathbf{k}_\pm = \pm\hat{\mathbf{x}}\,k)$ and FAI $n_s(\omega_s = i\gamma_s,\ \mathbf{k}_s = -\hat{\mathbf{x}}\,k)$, where $\omega_\pm = \omega = \omega_0 + i\gamma_s$.

Based on (7.4), the coupled mode equations for upper hybrid sidebands are

$$(\omega^2 + i\omega\nu_e - \omega_{uk}^2)k\phi_\pm = \pm\frac{i}{2}\omega_p^2\left(1 - \frac{|\Omega_e|}{\omega}\right)\left\langle\frac{E_{0\perp}n_{s\pm}}{n_0}\right\rangle \tag{7.18}$$

where the notations $n_{s+}^* = n_s = n_{s-}$ are used again; $\nu_{eL} = 0$.

The coupled mode equation for FAI is deduced from (7.5) to be

$$(\omega_s + i\nu_i)[\omega_s(\omega_s + i\nu_e + i\nu_i) - (k^2C_s^2 + |\Omega_e|\Omega_i)]\left(\frac{n_s}{n_0}\right)$$
$$= -\frac{i}{2}\frac{m_e}{m_i}\left(\frac{e}{m_e}\right)^2 \frac{k^3(\omega_u^2 - \omega^2)\Omega_e^2}{(\omega^2 - \Omega_e^2)(\omega + |\Omega_e|)\omega_p^2}(E_{0\perp}\phi_+^* + E_{0\perp}^*\phi_-) \tag{7.19}$$

where $\nu_{iL} = 0$ and $\nu_i = \nu_{in}$.

Equations (7.18) and (7.19) are analyzed in the same way as that for (7.9) and (7.10). The dispersion relation of upper hybrid OTSI is then derived to be

$$\left[\left(\gamma_1 + \frac{k^2C_s^2}{\omega_{LH}^2}\right)\left(\gamma_1 + 2\frac{m_e}{m_i} + \frac{k^2v_{te}^2}{\Omega_e^2}\right) + 2/3\frac{\gamma_1 k^2 v_{te}^2}{\Omega_e^2}\right]$$
$$\times\left[\Gamma^2 + \nu_e^2\frac{(\omega_0^2 + \Omega_e^2)^2}{\omega_0^2}\right]$$
$$= \frac{2}{3}\left[\frac{\omega_p}{\Omega_e(\omega_0 + \Omega_e)}\right]^2\left[\Gamma\left(1 - \frac{|\Omega_e|}{\omega_0}\right) - \nu_e^2\left(1 + \frac{\Omega_e^2}{\omega_0^2}\right)\right.$$
$$\left. - \frac{5}{4}\left(\frac{k\lambda_D}{\omega_0}\right)^2\left(1 + \frac{6m_e\Omega_e^2}{5m_i k^2 v_{te}^2}\right)(\omega_0^4 - \Omega_e^4)\right]\left(\frac{keE_{0\perp}}{m_e}\right)^2 \tag{7.20}$$

where $\gamma_1 = \gamma_s/\nu_e$, $k_D = \omega_p/v_{te}$ and $\Gamma = \omega_p^2 + 3k^2v_{te}^2 + \Omega_e^2 + \nu_{eh}^2 - \omega_0^2$.

Set $\gamma_1 = 0$ in (7.20), then the threshold field $|E_{0\perp}|_{th}$ of the instability is obtained to be

$$|E_{0\perp}|_{th} = \sqrt{3}\left[\frac{m_e\omega_0 C_s}{e(\omega_0 - |\Omega_e|)}\right]$$
$$\times\frac{\left\{\left(1 + \frac{k^2v_{te}^2}{2|\Omega_e|\Omega_i}\right)\left[\Gamma^2 + \nu_e^2\frac{(\omega_0^2+\Omega_e^2)^2}{\omega_0^2}\right]\right\}^{1/2}}{\left\{\left[\Gamma\left(1 - \frac{|\Omega_e|}{\omega_0}\right) - \left(1 + \frac{\Omega_e^2}{\omega_0^2}\right)[\nu_e^2 + \frac{3}{2}|\Omega_e|\Omega_i]\right\}^{1/2}} \tag{7.21}$$

The right-hand side of (7.21) has to be positive, it leads to the condition that

$$\Gamma > \left(1 + \frac{\Omega_e^2}{\omega_0^2}\right) \frac{[\nu_e^2 + \frac{3}{2}|\Omega_e|\Omega_i]\omega_0}{(\omega_0 - |\Omega_e|)} = a \tag{7.22}$$

The threshold field for $\Gamma = \Gamma_0 = a + [a^2 + \nu_e^2(\omega_0^2 + \Omega_e^2)^2/\omega_0^2]^{1/2}$ has the minimum value

$$|E_{uotsi}|_m = \sqrt{6}\left[\frac{m_e C_s \omega_0^{3/2}}{e(\omega_0 - |\Omega_e|)^{3/2}}\right]\left[\Gamma_0\left(1 + \frac{k^2 v_{te}^2}{2|\Omega_e|\Omega_i}\right)\right]^{1/2} \tag{7.23}$$

7.2.2 *Parametric decay instability — Decay of high-frequency heater wave to upper hybrid sideband $\phi(\omega, k)$ and lower hybrid wave $n_s(\omega_s, k_s)$*

For the upper hybrid sideband, the coupled mode equation (7.4) is reduced to

$$[-\Gamma + i\nu_e\omega]\phi = \frac{i}{2}\left(1 - \frac{|\Omega_e|}{\omega}\right)\left(\frac{\omega_p^2}{k}\right) E_{0\perp}\left(\frac{n_s^*}{n_0}\right) \tag{7.24}$$

where $\Gamma = \omega_{uk}^2 - \omega^2$, $\omega_{uk}^2 = \omega_k^2 + \Omega_e^2 + \nu_e^2$, and $\omega_k^2 = \omega_p^2 + 3k_\perp^2 v_{te}^2$.

For the lower hybrid decay mode, $\Omega_i^2 \ll |\omega_s|^2 \ll \Omega_e^2$, and $|\partial_t^2 \nabla_\perp^2|$ and $|\Omega_e^2 \nabla_z^2|$ are in the same order of magnitude. Thus, the coupled mode equation (7.5) is reduced to

$$\omega_s\left\{\omega_s\left[\omega_s + i\left(\nu_{es} + \frac{\nu_i}{\xi}\right)\right] - \omega_{LK}^2\right\}\left(\frac{n_s}{n_0}\right)$$
$$= i\frac{3}{2}\frac{m_e}{m_i}\left(\frac{e}{m_e}\right)^2 \frac{\Omega_e^2}{(\omega^2 - \Omega_e^2)(\omega + |\Omega_e|)}\left(\frac{k^5 v_{te}^2}{\omega_p^2}\right) E_{0\perp}\phi^* \tag{7.25}$$

where $k_{sx} \cong -k$ and $\Omega_e^2 \ll \omega_p^2$ are applied.

Equations (7.24) and (7.25) are combined to obtain the dispersion relation

$$[-\Gamma + i\nu_e\omega]\, \omega_s^* \left[\omega_s^{*2} - i\left(\nu_{es} + i\frac{\nu_i}{\xi}\right)\omega_s^* - \omega_{LK}^2\right] = \frac{3}{4}\frac{m_e}{m_i}\left(\frac{e}{m_e}\right)^2 \frac{\Omega_e^2}{(\omega + |\Omega_e|)^2}\left(\frac{k^4 v_{te}^2}{\omega}\right) E_{0\perp}^2 \tag{7.26}$$

where $\nu_i \ll \nu_{es}$ is applied.

We now set $\omega = \omega_{uk}$ and $\omega_s = \omega_{Lk}$ in (7.26), i.e., the growth rate $\gamma_k = 0$, the minimum threshold field $|E_{updi}|_m$ of the instability excited at the matching height of its upper hybrid sideband $\phi(k, \theta)$ is obtained to be

$$|E_{updi}|_m = \frac{1}{\sqrt{3}}\left(\frac{m_e}{e}\right)\left(1 + \frac{|\Omega_e|}{\omega_0}\right)[\nu_e(\xi\nu_{es} + \nu_i)]^{1/2}\left(\frac{\omega_0^2}{k^2 v_{te}}\right) \tag{7.27}$$

where $\nu_e = \nu_{ei} + \nu_{en}$, $\nu_{es} = \nu_e + \nu_{eLs}$, and $\nu_{eLs} = (\pi/2)^{1/2}(m_i/m_e)^{3/2} \times [\omega_{LH}^4\xi/k^3v_{te}^3(\xi - 1)^{1/2}]\exp[-m_i\omega_{LH}^2\xi/2m_ek^2v_{te}^2(\xi - 1)]$. The plasma frequency at the matching height is $\omega_p = (\omega_0^2 - \Omega_e^2 - 3k^2v_{te}^2)^{1/2} < (\omega_0^2 - \Omega_e^2)^{1/2}$. It indicates that the upper hybrid PDI prefers to be excited below the upper hybrid resonance layer.

The definitions of the symbols used in (7.12), (7.17), (7.23), and (7.27) and the results of the minimum threshold fields are summarized in the following: $C_s = [(T_e + 3T_i)/m_i]^{1/2}$ is the ion acoustic speed and $v_{te} = (T_e/m_e)^{1/2}$ is the electron thermal speed; ω_0, ω_{sr}, $\omega_{LH} = (|\Omega_e|\Omega_i)^{1/2}$, ω_p, and $|\Omega_{e,i}|$ are the heater wave, ion acoustic, lower hybrid, electron plasma, and electron and ion cyclotron radian frequencies; $\nu_{eh} = \nu_{en} + \nu_{ei} + \nu_{eL} = \nu_e + \nu_{eL}$, ν_{en} is electron–neutral elastic collision frequency, ν_{ei} is the electron–ion Coulomb collision frequency, $\nu_e = \nu_{en} + \nu_{ei}$, and $\nu_{eL} = (\pi/2)^{1/2}(\omega_0^2\omega_p^2/k_zk^2v_{te}^3)\exp(-\omega_0^2/2k_z^2v_{te}^2)$ is twice the electron Landau damping rate; $\nu_{es} = \nu_e + \nu_{eLs}$, and $\nu_{eLs} = (\pi/2)^{1/2}(m_i/m_e)^{3/2} [\omega_{LH}^4\xi/k^3v_{te}^3(\xi - 1)^{1/2}]\exp[-m_i\omega_{LH}^2\xi/2mk^2v_{te}^2(\xi - 1)]$ is twice the electron Landau damping rate on lower hybrid wave, where $\xi = 1 + (m_i/m_e)(k_z^2/k^2)$; $\nu_i = (\nu_{in} + \nu_{iL})$, ν_{in} is the ion–neutral collision frequency, $\nu_{iL} \cong (\pi/2)^{1/2}(\omega_s^2/k_z v_S)(T_e/T_i)^{3/2}\exp(-\omega_s^2/2k_z^2 v_{ti}^2)$ is twice the ion Landau damping rate on ion acoustic wave; ϑ is the oblique angle of the wavevector with respect to the magnetic field; $\Gamma_0 = a + [a^2 + \nu_e^2(\omega_0^2 + \Omega_e^2)^2/\omega_0^2]^{1/2}$ accounts for frequency mismatch, where $a = (1 + \Omega_e^2/\omega_0^2)\ [\nu_e^2 + (3/2)|\Omega_e|\Omega_i]/(1 - |\Omega_e|/\omega_0)$.

7.3 Correlation of Parametric Instabilities with the Observations

The results of analyses presented in Secs. 7.1 & 7.2 are applied to understand observations presented in Chapter 5. The relevant parametric values of the HF heating experiments conducted at Arecibo, Puerto Rico, Tromso, Norway, and Gakona, Alaska, are given in the following:

In Arecibo heating experiments, the parameters are:

$$\omega_0/2\pi = 5.1\,\mathrm{MHz}, |\Omega_e|/2\pi = 1.06\,\mathrm{MHz},$$

$$T_e = T_i = 1000\,\mathrm{K},$$

$$v_{te} \sim 1.23\times10^5\,\mathrm{m/s},\ v_{ti} = 7.17\times10^2\,\mathrm{m/s},\ C_s \sim 1.43\times10^3\,\mathrm{m/s},$$

$$\nu_{in} = 0.5\,\mathrm{s}^{-1},\ \nu_{iL} \sim 1.64\times10^3(k_s/k_R)\mathrm{s}^{-1},$$

$$\text{and}\quad \nu_e \cong \nu_{en} + \nu_{ei} \cong 500\,\mathrm{s}^{-1},$$

$$\text{and}\quad k_{\|0} \cong 4.377\pi\ (\text{i.e., } \lambda_{\|0} = 0.457\,\mathrm{m} = \lambda_R/2\sin\theta_m),$$

where k_s and k_R are the wavenumbers of ion acoustic wave and radar signal; $\lambda_R = 0.7\,\mathrm{m}$ is the wavelength of the 430 MHz radar signal and $\theta_m = 50°$ is the magnetic dip angle.

In Tromso heating experiments, the parameters are:

$$|\Omega_e|/2\pi = 1.35\,\mathrm{MHz},$$

$$T_e = 1500\,\mathrm{K},\quad T_i = 1000\,\mathrm{K},$$

$$v_{te} \sim 1.5\times10^5\,\mathrm{m/s},\quad v_{ti} = 7.17\times10^2\,\mathrm{m/s},$$

$$C_s \sim 1.52\times10^3\,\mathrm{m/s},$$

$$\nu_{in} = 0.8/0.5\,\mathrm{s}^{-1}\quad \text{and}\quad \nu_e = \nu_{en} + \nu_{ei} \sim 600\,\mathrm{s}^{-1},$$

$$\text{and} \quad \theta_m = 78°;$$

in the case of 933 MHz radar,

$$\omega_0/2\pi = 5.423, \quad k_{||1} = 12.17\pi \text{ (i.e., } \lambda_{||1} = 0.1644\,\text{m}),$$
$$\nu_{iL1} \sim 1.19 \times 10^4 (k_{s1}/k_{R1}) s^{-1}; \text{ and}$$

in the case of 224 MHz radar,

$$\omega_0/2\pi = 6.77\,\text{MHz}, \quad k_{||2} = 2.92\pi \text{ (i.e., } \lambda_{||2} = 0.685\,\text{m}),$$
$$\nu_{iL2} \sim 2.87 \times 10^3 (k_{s2}/k_{R2})\, s^{-1}.$$

In HAARP heating experiments, the parameters are:

$$\omega_0/2\pi = 5\,\text{MHz}, \quad |\Omega_e|/2\pi = 1.4\,\text{MHz},$$
$$T_e = 1500\,\text{K}, \quad T_i = 1000\,\text{K},$$
$$v_{te} \sim 1.5 \times 10^5\,\text{m/s}, \quad v_{ti} = 7.17 \times 10^2\,\text{m/s}, \quad C_s \sim 1.52 \times 10^3\,\text{m/s},$$
$$\nu_{in} = 0.5\,s^{-1}, \quad \nu_{iL} \sim 5.55 \times 10^3 (k_s/k_R) s^{-1},$$
$$\text{and} \quad \nu_e \cong \nu_{en} + \nu_{ei} \cong 600\,s^{-1},$$
$$\text{and} \quad k_{||0} \cong 5.77\pi \text{ (i.e., } \lambda_{||0} = 0.347\,\text{m} = \lambda_R/2\sin\theta_m),$$

where $\lambda_R = 0.67$ m is the wavelength of the 450 MHz radar signal and $\theta_m = 76°$.

Electron Landau damping rate ν_{eL} can be neglected in all cases of the heating sites and heater frequencies.

7.3.1 *Threshold fields of parametric decay instability and oscillating two-stream instability excited in Arecibo/Tromso/high frequency active auroral research program (HAARP) heating experiments*

We now use (7.12) and (7.17) to evaluate the threshold fields of OTSI and PDI which contribute to HFPLs; thus, $k_s = k = 2k_R$ and $\vartheta = \vartheta_0 = 90° - \theta_m$, where $2k_R = 5.71\pi$, $12.44\pi/2.99\pi$, and 6π for Arecibo's 430 MHz radar, EISCAT's 933/224 MHz radars, and HAARP's 450 MHz radar; $\vartheta_0 = 40°$, $12°$, and $14°$, at Arecibo, Tromso, and Gakona, respectively.

7.3.1.1 *Oscillating two-stream instability*

$$|E_{\text{otsi}}(2\mathrm{k_R},\theta_0)|_\mathrm{m} = \begin{cases} 0.46\,\mathrm{V/m} \text{ for } \mathrm{f_R} = 430\,\mathrm{MHz}, \text{ at Arecibo} \\ 0.44/0.49\,\mathrm{V/m} \text{ for } \mathrm{f_R/f_0} = 933/5.423 \text{ and} \\ \qquad 224/6.77\,\mathrm{MHz}, \text{ at Tromso} \\ 0.42\,\mathrm{V/m} \text{ for } \mathrm{f_R} = 450\,\mathrm{MHz}, \text{ at HAARP} \end{cases} \tag{7.28}$$

7.3.1.2 *Parametric decay instability*

$$|E_{\text{pdi}}(2\mathrm{k_R},\theta_0)|_\mathrm{m} = \begin{cases} 0.165\,\mathrm{V/m} & \text{at Arecibo} \\ 0.27/0.3\,\mathrm{V/m}, \text{ for } \mathrm{f_0} = & 5.423/6.77\,\mathrm{MHz} \\ & \text{at Tromso} \\ 0.24\,\mathrm{V/m} & \text{at HAARP} \end{cases} \tag{7.29}$$

The results presented in (7.28) and (7.29) indicate that the threshold fields of the OTSI are about 2.8, 1.6, and 1.7 times larger than those of PDI at the three sites exciting similar $(2\mathrm{k_R}, \vartheta_0)$ plasma lines in the corresponding nearby regions. Because these instabilities are excited very close to the HF reflection height, the RH circularly polarized HF heater wave is converted to the linearly polarized O-mode wave; in the absence of D region absorption, the electric field amplitude $\mathrm{E_{0z}}$ of the wave at a height h can be determined by the effective radiated power (ERP) of the HF transmitter through the relation $\mathrm{E_{0z}^2}/2\eta_0 = (\mathrm{ERP})/4\pi \mathrm{h}^2$, i.e., $\mathrm{E_{0z}} = [60 \times (\mathrm{ERP})]^{1/2}/\mathrm{h}$ V/m, where $\eta_0 = 120\pi$ Ohm is the intrinsic impedance of the free space. Moreover, in the region near the HF reflection height, the electric field amplitude of the heater is enhanced by a swelling factor of about 4 (see Sec. 4.2.2), where a factor of 2 is ascribed to the total reflection at cutoff and an additional factor of $\sim$2 ascribed to wave accumulation while slowing down the propagation to approach the turning point. Therefore, the required ERP of the HF transmitter to achieve $\mathrm{E_0} >$

0.5 V/m at 250 km height is less than 20 MW, which is much less than the available power of the transmitters at the three sites.

The threshold field/growth rate of the instability increases/decreases with the increase of the mismatch frequency $|\Delta\omega_1|$ of the sideband. Moreover, the OTSI and PDI contributing to HFPLs do not have the lowest threshold field and highest growth rate. Therefore, instabilities excited in regions outside of the matching heights are not likely to produce HFPLs. The excited Langmuir waves may cascade through a similar PDI.

7.3.2 *Threshold fields of upper hybrid parametric decay instability and oscillating two-stream instability excited in Arecibo/Tromso/HAARP heating experiments*

We now use (7.23) and (7.27) to evaluate the threshold fields of upper hybrid OTSI and PDI which are significant to the generation of FAIs, energetic electrons, and SEEs.

7.3.2.1 *Upper hybrid oscillating two-stream instability*

$$|E_{\mathrm{uotsi}}|_{\mathrm{m}} = \begin{cases} 4\left[1+\left(\dfrac{14}{\lambda}\right)^2\right]^{1/2} \mathrm{mV/m} & \text{at Arecibo} \\ 5.4/5.174 \times \left[1+\left(\dfrac{13.5}{\lambda}\right)^2\right]^{1/2} \mathrm{mV/m} & \\ \quad \text{for } \mathrm{f_0} = 5.423/6.77\,\mathrm{MHz} & \text{at Tromso} \\ 5.7\left[1+\left(\dfrac{13}{\lambda}\right)^2\right]^{1/2} \mathrm{mV/m} & \text{at HAARP} \end{cases} \tag{7.30}$$

For circularly polarized pump, $\mathrm{E}_{0\perp}^2/\eta_0 = (\mathrm{ERP})/4\pi \mathrm{h}^2$, i.e., $\mathrm{E}_{0\perp} = [30 \times (\mathrm{ERP})]^{1/2}/\mathrm{h}$ V/m. Neglecting D-region absorption and using 80 MW, 1 GW, and 1 GW ERP as the radiated powers at Arecibo, Tromso, and HAARP, $\mathrm{E}_{0\perp}$ at 250 km are estimated to be about 0.2, 0.69, and 0.69 V/m at three sites, respectively. Hence, these results presented in (7.30) show that upper hybrid OTSI can generate both large-scale and small-scale FAIs. In the large-scale case, i.e., $\lambda_1 \gg 14\,\text{m}$, the threshold field is very small and has a value of $(4 \sim 5.7)\,\text{mV/m}$.

7.3.2.2 *Upper hybrid parametric decay instability*

$$
|E_{\text{updi}}|_{\text{m}} = \begin{cases} 0.42\xi^{1/2}\lambda^2\left\{1 + 0.623\times 10^5\left[\dfrac{\xi}{(\xi-1)^{1/2}}\right]\lambda^3\exp\left[-\dfrac{37.13\lambda^2\xi}{(\xi-1)}\right]\right\}^{1/2} \\ \qquad \text{at Arecibo} \\ 0.482\xi^{1/2}\lambda^2\left\{1 + 0.752\times 10^5\left[\dfrac{\xi}{(\xi-1)^{1/2}}\right]\lambda^3\exp\left[-\dfrac{40.5\lambda^2\xi}{(\xi-1)}\right]\right\}^{1/2} \\ \qquad \text{for } f_0 = 5.423\,\text{MHz at Tromso and} \\ 0.721\xi^{1/2}\lambda^2\left\{1 + 0.752\times 10^5\left[\dfrac{\xi}{(\xi-1)^{1/2}}\right]\lambda^3\exp\left[-\dfrac{40.5\lambda^2\xi}{(\xi-1)}\right]\right\}^{1/2} \\ \qquad \text{for } f_0 = 6.77\,\text{MHz} \quad \text{at Tromso} \\ 0.42\xi^{1/2}\lambda^2\left\{1 + 0.87\times 10^5\left[\dfrac{\xi}{(\xi-1)^{1/2}}\right]\lambda^3\exp\left[-\dfrac{43.56\lambda^2\xi}{(\xi-1)}\right]\right\}^{1/2} \\ \qquad \text{at HAARP} \end{cases} \tag{7.31}
$$

The results of $|E_{\text{updi}}|_{\text{m}}$ presented in (7.31) are functions of λ, with ξ a variable parameter. These functions increase rapidly with the wavelength, i.e., $|E_{\text{updi}}|_{\text{m}} \propto \lambda^2$ for $\lambda_1 > 0.6\,\text{m}$. Moreover, the electron Landau damping on the lower hybrid waves imposes a lower bound on the wavelength of the instability. Thus, the wavelengths of the waves excited through upper hybrid PDI at Tromso and HAARP are limited in a range roughly between 0.25 to 1.5 m, and at Arecibo it is in a much narrower range around 0.36 m. In other words, the power density of the upper hybrid waves excited at Arecibo is expected to be much lower than those excited at Tromso and HAARP. This explains why SEEs are rarely observed in Arecibo heating experiments.

Problems

P7.1. In Arecibo HF heating experiment, O-mode HF heater of 5 MHz is transmitted and HFPLs are monitored by the 430 MHz UHF backscatter radar. If the background plasma has the electron temperature $T_e = 1200$ K and ion temperature $T_i = 800$ K, what are the frequencies of the HFPLs excited by the OTSI, PDI, and first Langmuir cascade.

P7.2. In HAARP HF heating experiment, O-mode HF heater of 3.2 MHz is transmitted.

(1) What is the plasma frequency at the upper hybrid resonance layer?

(2) What is the lower hybrid resonance frequency? (background magnetic field is about 0.5 Gauss)

(3) Upper hybrid wave and lower hybrid wave, with $k_{||}/k_{\perp} = 10^{-2}$, are excited parametrically by the O-mode heater; what are the frequencies of the excited upper hybrid wave and lower hybrid wave?

P7.3. Explain why the X-mode (LH circularly polarized) HF heater cannot excite parametric instabilities in the ionosphere.

P7.4. In high-latitude ionosphere, O-mode HF heater can excite upper hybrid wave via parametric instabilities in the upper hybrid resonance region. However, upper hybrid waves cannot be detected by the backscatter UHF radars. On the other hand, the generated upper hybrid waves can become pumps to parametrically excite Langmuir waves contributing to HFPLs, which are separated from those generated directly by the HF heater near the HF reflection height. Plot the wavevector matching and give the frequency matching condition.

Chapter 8

Nonlinear Plasma Waves

Parametric decay instability (PDI) and oscillating two-stream instability (OTSI) and the subsequent plasma wave cascades via follow-up parametric instability processes generate Langmuir waves and upper hybrid waves and broaden their frequency and wavenumber spectra. Although these instabilities can excite plasma waves with an angular distribution around the geomagnetic field, Langmuir waves propagating closely parallel to the geomagnetic field are dominant with the largest growth rates and upper hybrid waves are polarized closely perpendicular to the geomagnetic field to form standing waves. These dominant ones will grow to high intensity and aggregate together, the nonlinearity slows down wave dispersion. The self-induced density perturbations provide nonlinear feedback to modulate the envelopes of the aggregations; consequently, linear plasma waves evolve to nonlinear plasma waves.

Moreover, PDI-excited ion acoustic waves/lower hybrid waves are also likely to evolve into nonlinear ion waves. In low-frequency field, electrons tend to follow the ion oscillation to provide Debye shielding. In the linear regime, the ion density perturbation of the wave is small, quasi-neutrality is maintained. Consequently, ion acoustic wave is non-dispersive. As the wave grows to large amplitude in the nonlinear regime, Debye shielding on ion density perturbation of the wave becomes incomplete and is frequency dependent. A dispersion effect emerges to balance the plasma nonlinearity in the nonlinear evolution of the ion acoustic waves. Lower hybrid wave is dispersive due to the intrinsic lower hybrid resonance.

8.1 Formulation

8.1.1 *Nonlinear Schrödinger equation for Langmuir wave packet*

The coupled mode equation for the Langmuir wave is derived from (6.2) to (6.4), in which the magnetic field term (the second term on the LHS) of (6.3) is removed and (6.4) is rewritten as

$$\nabla \cdot \mathbf{E} = -\frac{e(n_e - \hat{n})}{\varepsilon_0} = -\frac{e\delta n_e}{\varepsilon_0} \tag{8.1}$$

where $n_e = n_0 + \delta n_e + n_s$; $\hat{n} = n_0 + n_s$; n_0, δn_e, and n_s are the unperturbed plasma density and electron density perturbations associated with Langmuir waves and low-frequency oscillations, respectively. Apply the operation $(\partial/\partial t + \nu_e)$ to (6.2), i.e., $(\partial/\partial t + \nu_e)$ (6.2), and with the aid of (6.3), we obtain

$$\begin{aligned}\frac{\partial}{\partial t}\left(\frac{\partial}{\partial t} + \nu_e\right)\delta n_e - \nabla \cdot (3v_{te}^2 \nabla \delta n_e) \\ = \langle \nabla \cdot (\nabla \cdot n_e \mathbf{v}_e \mathbf{v}_e) + (e/m_e)\nabla \cdot (n_e \mathbf{E})\rangle\end{aligned} \tag{8.2}$$

where $\langle\ \rangle$ stands for a bandpass filter, which keeps only terms in the same frequency range. Thus,

$$\langle \nabla \cdot (n_e \mathbf{E})\rangle = n_0 \nabla \cdot \mathbf{E} + \nabla \cdot (n_s \mathbf{E}) \tag{8.3}$$

Substitute (8.3) into (8.2) and with the aid of (8.1), an equation governing the evolution of the Langmuir wave field $\mathbf{E}$ is derived to be

$$\begin{aligned}\left[\frac{\partial}{\partial t}\left(\frac{\partial}{\partial t} + \nu_e\right) + \omega_p^2 - 3v_{te}^2 \nabla^2\right]\mathbf{E} \\ = -\frac{e}{\varepsilon_0}\langle \nabla \cdot (n_e \mathbf{v}_e \mathbf{v}_e)\rangle - \left(\frac{\omega_p^2}{n_0}\right)(n_s \mathbf{E})\end{aligned} \tag{8.4}$$

This is a nonlinear mode equation of the Langmuir wave. The nonlinear nature of the equation is shown implicitly by the two terms on the RHS of (8.4). The second term on the RHS of (8.4) depends explicitly on the wave-induced background density perturbation n_s, which modifies the dispersion property of the Langmuir wave self-consistently. The governing equation of this density perturbation is derived in the following.

We combine the momentum equations of electron and ion fluids by adding them together. The electric field terms and electron–ion collision terms in the two equations are cancelled; and the electron inertial term $m_e \partial \mathbf{v}_e/\partial t$ and the ion convective term $m_i \mathbf{v}_i \cdot \nabla \mathbf{v}_i$, which are small comparing to their respective counterpart, are neglected. The combined equation is obtained to be

$$m_i \left(\frac{\partial}{\partial t} + \nu_{in} \right) \mathbf{v}_i + m_e \mathbf{v}_e \cdot \nabla \mathbf{v}_e = -\frac{1}{n_0} \nabla (P_e + P_i) \tag{8.5}$$

Apply the quasi-neutrality and with the aid of the relations $\mathbf{v}_e \cdot \nabla \mathbf{v}_e = \nabla (v_e^2/2)$ and $\partial n_s/\partial t + \nabla \cdot (n_0 \mathbf{v}_i) = 0$ deduced from the continuity equation, (8.5) becomes

$$\left[\frac{\partial}{\partial t} \left(\frac{\partial}{\partial t} + \nu_{in} \right) - C_s^2 \nabla^2 \right] \left(\frac{n_s}{n_0} \right) = \frac{m_e}{m_i} \nabla^2 \left\langle \frac{v_e^2}{2} \right\rangle \tag{8.6}$$

where the implicit nonlinear term on the RHS of (8.6) is attributed to a ponderomotive force acting on the electron plasma. This force results from the non-uniform electron quiver motion in the Langmuir wave fields; as the electron plasma is pushed by this force, the induced electric field pulls ion plasma to move together. This term can be expressed explicitly in terms of the Langmuir wave field E.

8.1.2 *Ponderomotive force*

Consider the quiver motion of the electron fluid in a high-frequency electrostatic wave packet, the wave field is given to be $\mathbf{E}(\mathbf{z}, t) = \hat{\mathbf{z}} \int E_\omega(z, t) d\omega/2\pi$; the trajectory equations are

$$\frac{dz(t)}{dt} = v_z(t) \tag{8.7}$$

and

$$\frac{dv_z(t)}{dt} = -\frac{e}{m_e} E(z, t) = -\frac{e}{m_e} \int E_\omega[z(t), t] \frac{d\omega}{2\pi} \tag{8.8}$$

where v_z and z are the velocity and position of a layer in the electron fluid. The trajectory z(t) is expressed in the integration form to be

$$\begin{aligned} z(t) &= z_0 - \frac{e}{m_e}\int_0^t ds(t-s)\int E_\omega[z(s), s]\frac{d\omega}{2\pi} \\ &= z_0 + \Delta z(t) \end{aligned} \tag{8.9}$$

and

$$v_z(t) = -\frac{e}{m_e}\int_0^t ds\int E_\omega[z(s), s]\frac{d\omega}{2\pi} \tag{8.10}$$

where $z_0 = z(0)$ is the initial position of the layer. Take Tayler's series expansion,

$$\begin{aligned} E_\omega[z(t), t] &= E_\omega[z_0 + \Delta z(t), t] \\ &= E_\omega(z_0, t) + \Delta z(t)\frac{\partial}{\partial z}E_\omega(z_0, t) + \cdots \end{aligned} \tag{8.11}$$

Equation (8.10) becomes

$$\begin{aligned} v_z(t) &= -\frac{e}{m_e}\int_0^t ds\int\left[E_\omega(z_0, s) + \Delta z(s)\frac{\partial}{\partial z}E_\omega(z_0, s) + \cdots\right]\frac{d\omega}{2\pi} \\ &= v_{z0}(z_0, t) - \frac{e}{m_e}\int_0^t ds\left[\Delta z_0(z_0, s)\frac{\partial}{\partial z}\int E_\omega(z_0, s)\frac{d\omega}{2\pi}\right] + \cdots \\ &= v_{z0}(z_0, t) + \frac{e}{m_e}\int_0^t ds\left\{\frac{\partial}{\partial s}\Delta z_0(z_0, s)\right. \\ &\quad \times\left.\left[\int_0^s ds'\frac{\partial}{\partial z}\int E_\omega(z_0, s')\frac{d\omega}{2\pi}\right]\right\} + \cdots \end{aligned} \tag{8.12a}$$

where

$$v_{z0}(z_0, t) = -\frac{e}{m_e}\int_0^t ds\int E_\omega(z_0, s)\frac{d\omega}{2\pi}$$

and

$$\Delta z_0(z_0, t) = -\frac{e}{m_e}\int_0^t ds\left[(t-s)\int E_\omega(z_0, s)\frac{d\omega}{2\pi}\right]$$

With the aid of

$$\frac{\partial}{\partial s}\Delta z_0(z_0, s) = -\frac{e}{m_e}\int_0^s ds' \int [E_\omega(z_0, s')]\frac{d\omega}{2\pi} = v_{z0}(z_0, s),$$

(8.10a) becomes

$$v_z(t) = v_z(z_0, t) = v_{z0}(z_0, t) - \int_0^t ds \left[v_{z0}(z_0, s)\frac{\partial}{\partial z}v_{z0}(z_0, s)\right] + \cdots \tag{8.12b}$$

Finally, we obtain

$$m_e\frac{\partial}{\partial t}v_z(z_0, t) = -e\int E_\omega(z_0, t)\frac{d\omega}{2\pi} - m_e v_{z0}(z_0, t)\frac{\partial}{\partial z}v_{z0}(z_0, t) + \cdots \tag{8.13}$$

The first term on the RHS of (8.13) represents the local electric force on the electron fluid and the second term contains both low-frequency and high-frequency (harmonic) components. Let, $\langle\ \rangle$ represent a low-pass filter, $-m_e\langle v_{z0}(z_0, t)\partial/\partial z v_{z0}(z_0, t)\rangle = -m_e\partial/\partial z\langle v_{z0}^2/2\rangle \propto -\partial/\partial z\langle|\mathbf{E}(\mathbf{z}, t)|^2/2\rangle$ from the second term is a "ponderomotive force" induced by the Langmuir wave packet, impelling the electron fluid.

This ponderomotive force drives non-oscillatory density perturbation n_s as illustrated in Fig. 8.1, thus $|\partial/\partial t(\partial/\partial t + \nu_{in})(n_s/n_0)| \ll |C_s^2\nabla^2(n_s/n_0)|$ and (8.6) is approximated to obtain $(n_s/n_0) \cong -(m_e/m_i)\langle(v_e^2/2)\rangle/C_s^2$. Moreover, we will assume that $|\langle(\nabla\cdot(n_e\mathbf{v}_e\mathbf{v}_e)\rangle| \ll |(\omega_p^2/4\pi e n_0)(n_s\mathbf{E})|$ and $T_{eff} \cong T_e$ in (4.43), i.e., the time average

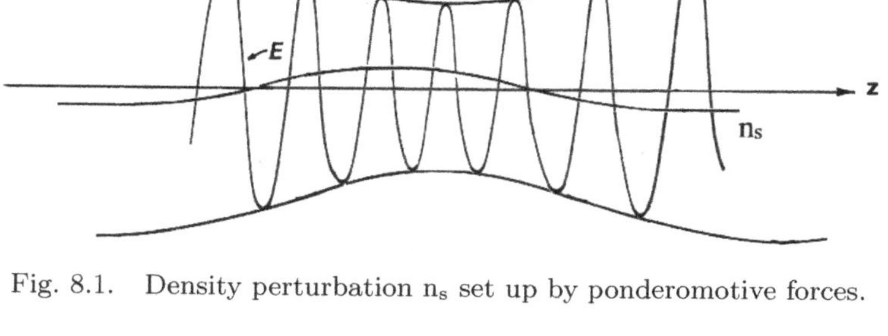

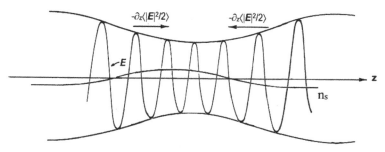

Fig. 8.1. Density perturbation n_s set up by ponderomotive forces.

wave energy per electron $T_w \ll T_e$. These assumptions impose an upper bound on the amplitude and lower bound on the scale length of a nonlinear Langmuir wave governed validly by the equation to be derived. Thus, (8.4) is reduced to

$$\left[\frac{\partial}{\partial t}\left(\frac{\partial}{\partial t}+\nu_e\right)+\omega_p^2-3v_{te}^2\nabla^2\right]E=\left(\frac{\omega_{pi}^2}{C_s^2}\right)\left\langle\frac{v_e^2}{2}\right\rangle E \quad (8.14)$$

Set $E = \mathcal{E}(\xi,\tau)\exp[-i(\omega_1 t - k_1 z)] + c.c.$, where $\xi = k_1 z - \tau$, $\tau = k_1 v_g t$, $v_g = \partial\omega/\partial k|_{\omega_1,k_1} = 3k_1 v_{te}^2/\omega_1$, and $\mathcal{E}(\xi,\tau)$ is the envelope of the wave, ω_1 and k_1 are the carrier frequency and wavenumber; thus, $v_e \cong -i[e\mathcal{E}(\xi,\tau)/m_e\omega_1]\exp[-i(\omega_1 t - k_1 z)] + c.c.$. Take forward wave approximation (i.e., neglect $\partial^2/\partial\tau^2$ term) and set $a(\xi,\tau) = e\mathcal{E}(\xi,\tau)/m_e\omega_1 v_{te}$, (8.14) reduces to

$$-\frac{1}{2}\frac{\partial^2}{\partial\xi^2}a+(V_l-A|a|^2)a=i\frac{\partial}{\partial\tau}a \quad (8.15)$$

where $V_l = -\Delta\omega^2/6k_1^2v_{te}^2$, $\Delta\omega^2 = \omega^2 - (\omega_p^2 + 3k^2v_{te}^2)$, and $A = (\omega_{pi}^2/6k_1^2C_s^2)$. Equation (8.15) is a nonlinear Schrödinger equation, where the wave function $a(\xi,\tau)$ represents the envelope of a Langmuir wave packet.

8.1.3 *Korteweg–de Vries (KdV) equation for ion acoustic waves*

Fluid equations of unmagnetized plasma in collisionless case are used to describe parallel propagating ion waves. Without the assumption of quasi-neutrality, the Possion's equation is included to relate the wave field to the density oscillations. Add the electron and ion momentum equations and neglect the electron inertial terms, a one fluid equation is obtained to be

$$\left[\frac{\partial}{\partial t}V_{si}+\frac{\partial}{\partial z}\left(\frac{V_{si}^2}{2}\right)\right]=-C_s^2\frac{\partial}{\partial z}\left(\frac{\delta n_{si}}{n_0}\right)-\left(\frac{T_e}{n_0m_i}\right)\frac{\partial}{\partial z}(\delta n_{se}-\delta n_{si}) \quad (8.16)$$

From the electron momentum equation (with the inertial terms neglected), it gives $E \sim -(T_e/n_0e)\partial\delta n_{se}/\partial z \sim -(T_e/n_0e)\partial\delta n_{si}/\partial z$.

With the aid of the Possion's equation $(\delta n_{se} - \delta n_{si}) = -(\varepsilon_0/e)\partial E/\partial z \sim (v_s/\omega_{pi})^2 \partial^2 \delta n_{si}/\partial z^2$, (8.16) becomes

$$\left[\frac{\partial}{\partial t} V_{si} + \frac{\partial}{\partial z}\left(\frac{V_{si}^2}{2}\right)\right] = -C_s^2 \frac{\partial}{\partial z}\left(\frac{\delta n_{si}}{n_0}\right) - \left(\frac{v_s^4}{\omega_{pi}^2}\right)\frac{\partial^3}{\partial z^3}\left(\frac{\delta n_{si}}{n_0}\right) \tag{8.17}$$

where $v_s^2 = T_e/m_i$ and $\omega_{pi}^2/v_s^2 = k_{De}^2$. We next take the partial time derivative operation $\partial/\partial t$ on both sides of (8.17); with the aid of the ion continuity equation $\partial/\partial t(\delta n_{si}/n_0) = -\partial V_{si}/\partial z$, it results to

$$\frac{\partial}{\partial t}\left[\frac{\partial}{\partial t} V_{si} + \frac{\partial}{\partial z}\left(\frac{V_{si}^2}{2}\right)\right] = C_s^2 \frac{\partial^2}{\partial z^2} V_{si} + \left(\frac{v_s^4}{\omega_{pi}^2}\right)\frac{\partial^4}{\partial z^4} V_{si} \tag{8.18}$$

In the linear regime, the second (nonlinear) term in the LHS bracket and the second (dispersion) term on the RHS of (8.18) are neglected; (8.18) becomes $(\partial^2/\partial t^2 - C_s^2\partial^2/\partial z^2)V_{si} = 0$, determining the linear dispersion relation $\omega_s = kC_s$ of the ion acoustic wave.

Equation (8.18) is transformed to a moving frame at the velocity C_s by setting $t' = t$ and $z' = z - C_s t$ and letting $V_{si}(z,t) = V_s(z',t')$; thus $\partial/\partial t \rightarrow \partial/\partial t' - C_s\partial/\partial z'$ and $\partial/\partial z \rightarrow \partial/\partial z'$, and (8.18) becomes

$$\frac{\partial}{\partial t'}\left[\frac{\partial}{\partial t'} V_s + \frac{\partial}{\partial z'}\left(\frac{V_s^2}{2}\right)\right] - 2C_s \frac{\partial}{\partial z'}\left[\frac{\partial}{\partial t'} V_s + \frac{\partial}{\partial z'}\left(\frac{V_s^2}{4}\right)\right] = \left(\frac{v_s^4}{\omega_{pi}^2}\right)\frac{\partial^4}{\partial z'^4} V_s \tag{8.19}$$

In the moving frame, the frequencies of the linear ion acoustic waves are downshifted to zero, thus, $|\partial/\partial t'| \ll |C_s\partial/\partial z'|$ and the first term on the LHS of (8.19) is neglected. Introducing the dimensionless variables and function: $\tau = \omega_{pi}t', \eta = Kz'$, and $\phi(\eta,\tau) = \alpha V_s$, where $K = (2C_s\omega_{pi}^3/v_s^4)^{1/3} = (2C_s/v_s)^{1/3}k_{De}$ and $\alpha = K/12\omega_{pi} = (12v_s)^{-1}(2C_s/v_s)^{1/3}$, (8.19) is converted to a standard KdV equation

$$\frac{\partial}{\partial \tau}\phi + \frac{\partial^3}{\partial \eta^3}\phi + 6\phi\frac{\partial}{\partial \eta}\phi = 0 \tag{8.20}$$

8.2 Analysis

8.2.1 *Nonlinear Langmuir wave packet*

The second term on the LHS of (8.15) contains a cubic nonlinear term $-A|a|^2 a$; and in the Hamiltonian representation, the coefficient $V_l - A|a|^2$ of the second term represents a potential operator of the system, where V_l is a linear potential function and $-A|a|^2$ is a self-induced nonlinear potential function. The total potential function $(V_l - A|a|^2)$ varies with the intensity of the wave function, thus it is possible to trap the wave function in self-induced potential well. When this occurs, the packet of Langmuir waves evolves into a localized nonlinear wave having an envelope called "Soliton".

Substituting $a(\xi, \tau) = F(\xi)e^{-i(\varpi\tau+\Phi)}$ into (8.15), where $F(\xi)$ is a real function, yields

$$F'' + \Lambda F + 2AF^3 = 0 \tag{8.21}$$

where $F'' = d^2F/d\xi^2$ and $\Lambda = 2(\varpi - V_l)$.

Consider F as the spatial coordinate of a unit mass object in a one-dimensional space, (8.21) represents an equation of motion of the object, which is accelerated by a force equal to $-(\Lambda F + 2AF^3)$. This equation is integrated to be

$$(1/2)(F')^2 + (1/2)(\Lambda F^2 + AF^4) = C_1 \tag{8.22}$$

where $F' = dF/d\xi$ and the integration constant $C_1 = (1/2)[F'^2(0) + \Lambda F^2(0) + AF^4(0)]$.

Equation (8.22) is the energy conservation equation describing the trajectory of the object in the potential field $VE = (1/2)(\Lambda F^2 + AF^4) = TE-KE$, where the total energy $TE = C_1$ and kinetic energy $KE = (1/2)(F')^2$. Two typical plots of the potential function in the cases of $\Lambda > 0$ and $\Lambda < 0$ are illustrated in Fig. 8.2(a). As shown, both plots represent potential wells.

In the case of $\Lambda > 0$, the total energy C_1 of the trapped object is larger than zero, i.e., $C_1 > 0$; the trapped object is bounced back and forth in the potential well to have a periodic trajectory $F_{p1}(\xi)$, which is a symmetric alternate function illustrated in Fig. 8.2(b).

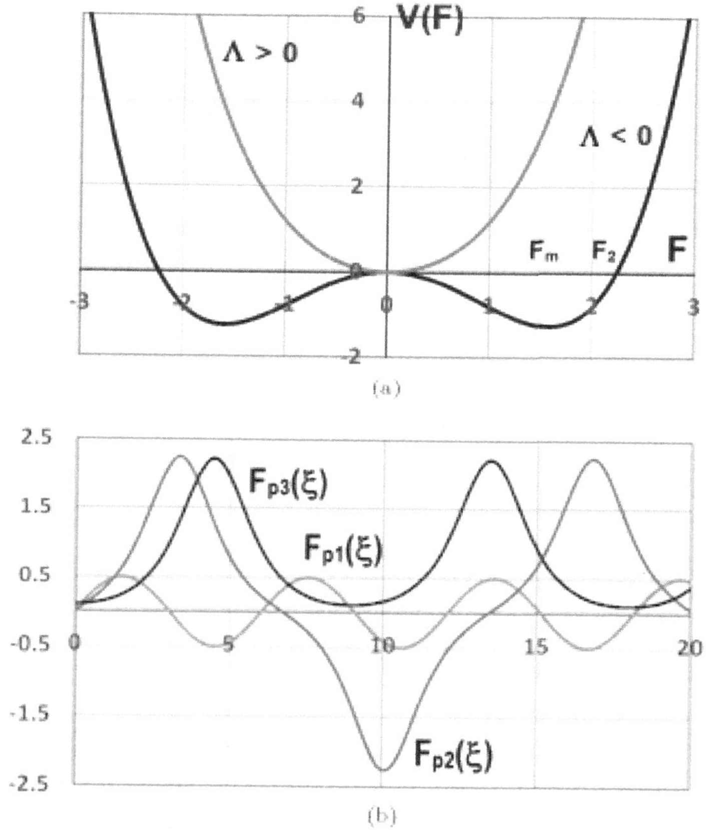

Fig. 8.2. (a) Two potential distributions presenting three type potential wells and (b) the back and forth trajectories trapped in the three different potential wells shown in (a).

In the case of $\Lambda < 0$, the trapped object has $C_1 > 0$ or $(1/2)(\Lambda F_m^2 + AF_m^4) < C_1 < 0$. If $C_1 > 0$, the bounce motion of the object has a periodic trajectory $F_{p2}(\xi)$, which is a symmetric alternate function illustrated in Fig. 8.2(b). If $(1/2)(\Lambda F_m^2 + AF_m^4) < C_1 < 0$, the periodic trajectory $F_{p3}(\xi)$ of the object is a non-alternate function as shown in Fig. 8.2(b). Moreover, there is also exists a non-periodic trajectory if $C_1 = 0$ in the case of $\Lambda < 0$. Let the object start at $F = F_1(= 0)$ and examine the motion of the object in the region between F_1 and F_2 in

Fig. 8.2(a). Initially, it moves very slowly to the right. As it drops into the potential well, it moves quickly toward the potential minimum at F_m. After passing the potential minimum, the object starts to climb up to the turning point at $F = F_2$, where the object is bounced back into the potential well. It quickly passes the potential minimum and then climbs up toward the starting point F_1. It takes a long time for the object to reach F_1, where the object stays. This represents a solitary trajectory to be shown in the following.

8.2.1.1 *Periodic solutions*

As illustrated in Fig. 8.2(b), the solution of (8.22) for $C_1 \neq 0$ is periodic. The analytical solutions of (8.21) are found in special cases as illustrated in the following.

(1) For $\varpi > 0$.

Let $\eta_1 = (2\varpi)^{1/2}\xi$ and $F(\xi) = F_1 y(\eta_1)$, (8.21) is normalized to be

$$y'' + \left(1 - \frac{V_1}{\varpi}\right) y + \left(1 - \frac{AF_1^2}{\varpi}\right) y^3 = 0 \tag{8.23}$$

where $y'' = d^2y/d\eta_1^2$. For $V_l = AF_1^2$, i.e., $F_1^2 = -(\Delta\omega^2 C_s/\omega_{pi}^2 v_{te}^2)$, the solution of (8.23) is a Jacobi elliptic function $cn(\eta_1, \gamma_1)$, where $\gamma_1 = (V_l/2\varpi)^{1/2}$. Because $F_1^2 > 0$, it requires that $\Delta\omega^2 < 0$; namely, the nonlinear periodic solution exists in the region above the matching height.

(2) For $\varpi < 0$ and $\Delta\omega^2 > 0$.

Let $\eta_2 = (2|V_l|)^{1/2}\xi$ and $F(\xi) = F_2 y(\eta_2)$, (8.21) is normalized to be

$$y'' + \left(1 - \frac{|\varpi|}{|V_1|}\right) y + \left(\frac{AF_2^2}{|V_1|}\right) y^3 = 0 \tag{8.24}$$

where $y'' = d^2y/d\eta_2^2$. For $|\varpi| = AF_2^2$, i.e., $F_2^2 = (6|\varpi|k_1^2 C_s^2/\omega_{pi}^2)$, the solution of (8.24) is a Jacobi elliptic function $cn(\eta_2, \gamma_2)$, where $\gamma_2 = (|\varpi/2V_1|)^{1/2}$.

(3) For $\varpi = 0$ and $\Delta\omega^2 < 0$.

Let $\eta_3 = (V_l)^{1/2}\xi$ and $F(\xi) = F_3 y(\eta_3)$, (8.21) is normalized to be

$$y'' - 2y + \left(\frac{2AF_3^2}{V_l}\right) y^3 = 0 \tag{8.25}$$

where $y'' = d^2y/d\eta_3^2$. For $V_l = AF_3^2$, i.e., $F_3^2 = -(\Delta\omega^2 C_s/\omega_{pi}^2 v_{te}^2)$, the solution of (8.25) is a Jacobi elliptic function $dn\ (\eta_3, 0)$.

8.2.1.2 *Solitary solution — Langmuir soliton*

Consider a localized solution of (8.21), it requires that $F = 0 = F'$ as $|\xi| \to \infty$. Thus, $C_1 = 0$, and (8.22) is re-expressed to be

$$F'^2 - \alpha F^2 + AF^4 = 0 \tag{8.26}$$

where $\alpha = -\Lambda = 2(V_l - \varpi)$. A solitary solution of (8.26) is given by

$$F(\xi) = \left(\frac{\alpha}{A}\right)^{1/2} \operatorname{sech}\sqrt{\alpha}\xi \tag{8.27}$$

subject to $\alpha > 0$ i.e., the wave energy ϖ is less than the linear potential energy V_l; it requires that $\Delta\omega^2 < -6\varpi k_1^2 v_{te}^2$, in the region above the matching height as well as higher than the region appearing the nonlinear periodic wave $cn(\eta_1, \gamma_1)$. (8.27) indicates that the width of a Langmuir soliton is inversely proportional to its amplitude. In OTSI, $\Delta\omega^2 = -\omega_1\nu_e$. Consider the case of $\varpi = 0$, and with the aid of (8.27) the Langmuir soliton field is obtained to be

$$E(z,t) = 2\left(\frac{2\omega_1\nu_e}{\omega_{pi}^2}\right)^{1/2}\left(\frac{m_e\omega_1 C_s}{e}\right) \times \operatorname{sech}\left[\left(\frac{\omega_1\nu_e}{\omega_{pi}^2}\right)^{1/2}(z - v_g t)\right]\cos(\omega_1 t - k_1 z + \Phi) \tag{8.28}$$

This envelope soliton (8.28) and the self-induced potential well ($\propto -F^2$) to trap this solitary wave are plotted in Fig. 8.3. Soliton is the result of the balance between the dispersion effect (represented by

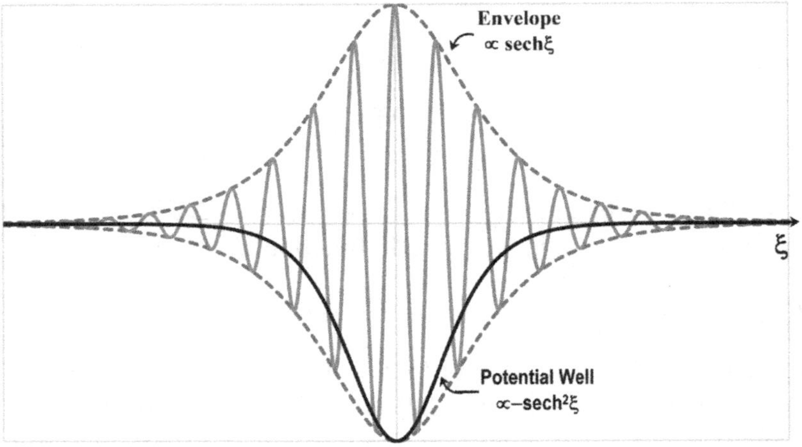

Fig. 8.3. Envelope soliton solution of the nonlinear Schrödinger Equation (8.15) and the self-induced potential well trapping the soliton.

the first term on the LHS of (8.15)) and the nonlinearity (represented by the third term on the LHS of (8.15)) of the medium. The nonlinearity of the medium focuses wave to overcome wave dispersion in the propagation, thus a shape-preserved solitary wave can exist.

With the aid of the relations $\int_{-\infty}^{\infty}$ sechz dz $= \pi$ and $\int_{-\infty}^{\infty}$ sech^{2}z dz $= 2$, the area $\int_{-\infty}^{\infty} \mathrm{F}(\xi)\mathrm{d}\xi = \pi/\mathrm{A}^{1/2}$ and energy $\int_{-\infty}^{\infty} \mathrm{F}^2(\xi)\mathrm{d}\xi = 2\alpha^{1/2}/\mathrm{A}$ of a Langmuir soliton are found to be a constant and to be proportional to the amplitude $\propto \alpha^{1/2}$, respectively. Because soliton is a localized entity, several solitons may appear simultaneously and interact each other in the transient period.

8.2.2 *Nonlinear ion acoustic waves*

KdV equation (8.20) has a solitary solution of the form

$$\phi(\eta,\tau) = \phi(\eta - \mathrm{A_s}\tau) = (1/2)\mathrm{A_s}\,\mathrm{sech}^2[(1/2)\sqrt{\mathrm{A_s}}(\eta - \mathrm{A_s}\tau)] \qquad (8.29)$$

It is shown that the velocity (in the moving frame) of an ion acoustic soliton is proportional to its amplitude $\mathrm{A_s}$ and the width is inversely proportional to the square root of the amplitude. The area

$\int_{-\infty}^{\infty} \phi(\eta,\tau)d\eta = 2A_s^{1/2}$ and energy $\int_{-\infty}^{\infty} \phi^2(\eta,\tau)d\eta = (2/3)A_s^{3/2}$ of an ion acoustic soliton are amplitude dependent. These relations impose conditions on the source (initial) pulse which is likely to evolve nonlinearly to become a soliton.

8.2.2.1 *Analysis*

Consider a more general solution of the form $\phi(\eta,\tau) = f(\xi)$, where $\xi = \eta - A_s\tau$, (8.20) becomes

$$\frac{d^3f}{d\xi^3} - A_s\frac{df}{d\xi} + 3\frac{df^2}{d\xi} = 0 \qquad (8.30)$$

This equation is integrated to be

$$(1/2)f'^2 + (-C_0f - A_sf^2/2 + f^3) = C_1 \qquad (8.31)$$

where $f' = df/d\xi$ and the integration constants $C_0 = [f''(0) - A_sf(0) + 3f^2(0)]$ and $C_1 = (1/2)[f'^2(0) - 2C_0f(0) - A_sf^2(0) + 2f^3(0)]$. Again, set f as the spatial coordinate of a unit mass object in one-dimensional space, (8.31) represents the energy conservation equation describing the trajectory of the object in the potential field VE $= (-C_0f - (1/2)A_sf^2 + f^3) =$ TE $-$ KE, where the total energy TE $= C_1$ and kinetic energy KE $= (1/2)(f')^2$.

The typical plots $V_1(f)$ and $V_0(f)$ of the potential field VE in the cases of $C_0 > 0$ and $C_0 = 0$ are illustrated in Fig. 8.4(a).

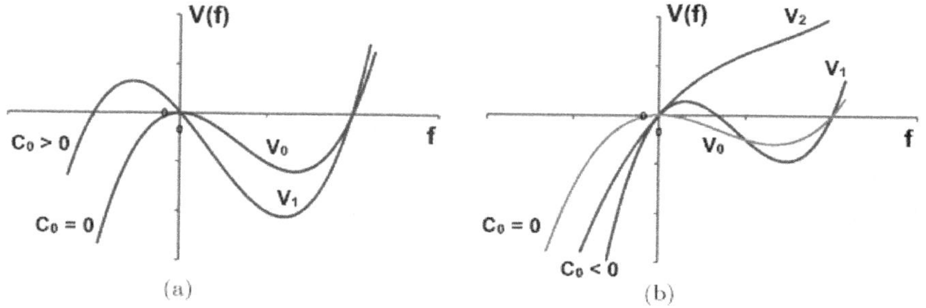

Fig. 8.4. (a) Two typical plots of the potential functions V_1 and V_0 for $C_0 > 0$ and $C_0 = 0$ and (b) two different type potential functions V_1 and V_2 for $C_0 < 0$. V_0 for $C_0 = 0$ in (a) and (b) are the same.

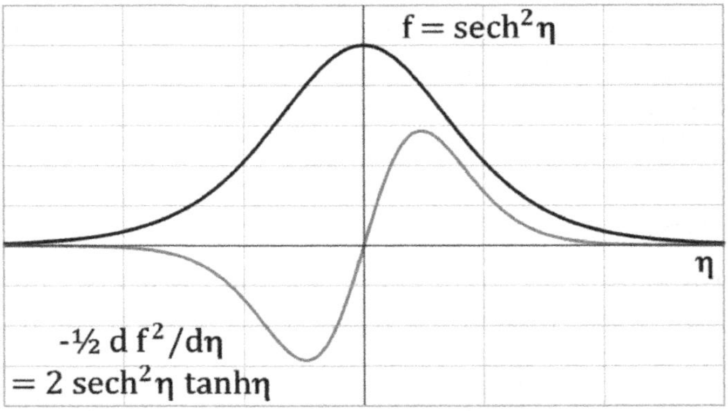

Fig. 8.5. Ion acoustic soliton and the self-induced ponderomotive force to balance the dispersion.

Both plots exhibit a potential well, which traps objects to force periodic motion. In the case of $C_0 < 0$, two typical plots $V_1(f)$ and $V_2(f)$ of the potential field VE are presented in Fig. 8.4(b), where $V_0(f)$ of Fig. 8.4(a) is also plotted for comparison. It is noted that the vertical scale of Fig. 8.4(b) is larger than that of Fig. 8.4(a). The potential function V_2 in Fig. 8.4(b) cannot trap the object; thus, the trajectory of the object in this potential field is not converted to a physical solution. On the other hand, the potential functions V_0 and V_1 in Figs. 8.4(a) and 8.4(b) can trap the object to give a periodic trajectory, which represents a periodic solution of (8.30).

8.2.2.2 *Soliton trapped in self-induced potential well*

With the solitary solution (8.29), $C_0 = C_1 = 0$ are obtained. Thus, only the potential distribution $V_0(f)$ in Fig. 8.4 can maintain a solitary trajectory, which converts to a soliton as shown in Fig. 8.5, in which the self-induced ponderomotive force to balance the wave dispersion is also plotted.

An ion acoustic soliton can be seen as a traveling plasma density bump, which is converted from (8.29) through the ion continuity equation $\partial/\partial t(n_{si}/n_0) + \partial V_{si}/\partial z = 0$, where $V_{si} = \phi/\alpha$, and $\tau = \omega_{pi}t$

and $\eta = K(z - C_s t)$, to be

$$\frac{n_{si}}{n_0} = 3A'_s \text{sech}^2 \left[\left(\frac{C_s}{v_s} \right) \sqrt{A'_s} k_{De}(z - C'_s t) \right] \tag{8.32}$$

where $A'_s = (v_s/2C_s)^{4/3} A_s$ and $C'_s = (1 + 2A'_s)C_s$.

It is shown that plasma can support solitons, which are shape-preserved localized wave packets. However, it is realized that soliton is not a necessity of the plasma nonlinearity, the nonlinear plasma waves, in general, are periodic; soliton(s) may appear when the source wave function has a localized form. In heating experiments, heating wave covers a large cross-section area (a few tens of km in diameter) and the Langmuir sidebands in OTSI are mainly standing waves. Therefore, under the normal condition, it is not likely to excite localized plasma waves as the source waves to evolve nonlinearly into solitons.

Problems

P.8.1. Derive the nonlinear equation for ion waves interacting with high-frequency electron plasma $\alpha\alpha$ waves and show that the result is (8.6)

P.8.2. Derive the ponderomotive force acting on an electron in a uniform magnetic field subject to a pump wave and low-frequency decay products.

P.8.3. Show that the solitary function (8.27) is a solution of (8.29). Calculate the area and energy of a Langmuir soliton.

P.8.4. Show that the solitary function (8.29) is a solution of the KdV equation (8.20). Calculate the area and energy of an ion acoustic soliton.

Chapter 9

Ionospheric Very- Low-Frequency Transmitter

9.1 Background

Communication to submerged submarines dictates the VLF frequency band. This is because sea water skin depth $d_s = (\pi f \mu_0 \sigma)^{-1/2}$ is frequency dependent; the optimal frequency for an electromagnetic wave to penetrate to a depth h (in meters) into the sea water is given by $f_m \sim (16/h^2)$ kHz, which is in the VLF band.

A small linear dipole of size L has a radiation resistance $R_r = 20(\pi L/\lambda)^4$ ohm, where λ is the wavelength. Let, R_a be the resistance of the antenna, where $R_a = \rho_a L/a_a$ with ρ_a and a_a are the resistivity and cross-section area of the antenna, the input resistance $R_{in} = R_r + R_a$. For a fixed radiation power $P_r = (1/2)\, I_0^2 R_r$, the input current $I_0 \propto (L/\lambda)^{-2}$. The ohmic loss in the antenna $P_\ell = (1/2)\, I_0^2 R_a$, is proportional to $(L/\lambda)^{-3}$. A VLF communications system requires a very large-size antenna for effective radiation. Moreover, VLF waves are ducted by the earth–ionosphere waveguide to polarize vertically; such a polization oughts to be transmitted by a vertical antenna, which has a limitation on its physical height L. In other words, a practical ground-based vertical VLF antenna is electrically too short, i.e., $L/\lambda \ll 1$, to be an effective transmitter. Transmitting VLF waves in the ionosphere has the advantage on the physical scale of an ionospheric virtual antenna.

Two approaches are studied. the first, classic approach employs a ground-based HF heater to directly modulate ionospheric electrojet, which appears in the D and lower E regions of high-latitude ionosphere and has a temperature-dependent electric conductivity. The intensity-modulated HF heater induces an alternate current in the electrojet, which serves as a virtual antenna to transmit VLF waves. The spatial and temporal variations of the electrojet impact the reliability of the classic approach. The second, beat-wave approach also employs a ground-based HF heater; however, in this approach, the heater operates in a continuous wave mode at two HF frequencies separated by the desired VLF frequency. the VLF radiation intensity depends upon the HF heater intensity rather than the electrojet strength.

9.2 Very- Low-Frequency Radiation of Modulated Ionospheric Electrojet

9.2.1 *Formulation of electrojet modulation*

The polar electrojet, consisting of Pederson and Hall current, is driven by a DC electric field, set to be $\mathbf{E}_0 = \hat{\mathbf{x}}E_0$, in the D and lower E regions of the ionosphere, which is embedded in the geomagnetic field with the magnetic induction $\mathbf{B}_o = -\hat{\mathbf{z}}B_0$. The Pederson current is in the same direction as the driving field (in the x-direction) and the Hall current in the y-direction. The conductivity $\sigma = n_0 e^2 \nu_{en}/m_e(\nu_{en}^2 + \Omega_e^2)$ of the electron plasma in both current components vary with the collision frequency ν_{en}, which is electron temperature dependent, i.e., $\nu_{en} \propto T_e^{5/6}$. Thus, the electrojet current can be modulated through a time varying heating of the electron plasma. The electron drift velocity is given to be

$$\mathbf{u}_e = -\frac{e}{m_e}E_0\frac{(\hat{\mathbf{x}}\nu_{en} - \hat{\mathbf{y}}\Omega_e)}{(\nu_{en}^2 + \Omega_e^2)} \tag{9.1}$$

Then, the electron electrojet current density is obtained to be

$$\mathbf{J}_e = -en_0\mathbf{u}_e = \left(\hat{\mathbf{x}} - \hat{\mathbf{y}}\frac{|\Omega_e|}{\nu_{en}}\right)\sigma E_0 \tag{9.2}$$

The modulated electrojet current is modeled, as a localized transverse current source positioned at $\mathbf{r}' = 0$, to be

$$\mathbf{J}_e(\mathbf{r}', t) = V\left(\frac{n_0 e^2}{m_e}\right)\left\{\frac{[\hat{\mathbf{x}}\nu_{en}(t) - \hat{\mathbf{y}}|\Omega_e|]}{[\nu_{en}(t)^2 + \Omega_e^2]}\right\} E_0 \delta(\mathbf{r}') \tag{9.3}$$

where V is the volume of the current source.

The vector potential $\mathbf{A}(\mathbf{r}, t)$ satisfies the inhomogeneous wave equation

$$\nabla^2 \mathbf{A} - \frac{1}{c^2}\frac{\partial^2}{\partial t^2}\mathbf{A} = -\mu_0(\mathbf{J} + \mathbf{J}_e) \tag{9.4}$$

under the transverse gauge condition $\nabla \cdot \mathbf{A} = 0$; $\mathbf{J}$ is the current density induced by the radiation field $\mathbf{E} = -\partial\mathbf{A}/\partial t$ in background plasma, which is derived to be $\mathbf{J} \cong i\varepsilon_0(\omega_p^2/|\Omega_e|)\partial\mathbf{A}/\partial t$; thus, (9.4) becomes

$$\nabla^2 \mathbf{A} - \frac{1}{c^2}\frac{\partial^2}{\partial t^2}\mathbf{A} + i\frac{1}{c^2}\left(\frac{\omega_p^2}{|\Omega_e|}\right)\frac{\partial}{\partial t}\mathbf{A} = -\mu_0 \mathbf{J}_e \tag{9.5}$$

which has the solution

$$\begin{aligned}\mathbf{A}(\mathbf{r}, t) &= \left(\frac{\mu_0}{4\pi}\right)\int d\mathbf{r}' \int dt' \left[\frac{\mathbf{J}_e(\mathbf{r}', t)}{|\mathbf{r} - \mathbf{r}'|}\right]\delta\left(t' + \frac{|\mathbf{r} - \mathbf{r}'|}{v_A} - t\right) \\ &\cong V\left(\frac{\mu_0 n_0 e^2 E_0}{4\pi m_e}\right)\left\{\frac{[\hat{\mathbf{x}}\nu_{en}(\tau_r) - \hat{\mathbf{y}}|\Omega_e|]}{[\nu_{en}^2(\tau_r) - \Omega_e^2]}\right\}\frac{1}{r} = \frac{\mathbf{p}(\mathbf{r}, t)}{r}\end{aligned} \tag{9.6}$$

where $\tau_r = t - r/v_A$ is a retarded time, $\mathbf{p}(\mathbf{r}, t) = (\mu_0 V n_0 e^2 E_0/4\pi m_e)$ $\{[\hat{\mathbf{x}}\nu_{en}(t - r/v_A) - \hat{\mathbf{y}}|\Omega_e|]/[v_{en}^2(\tau_r) + \Omega_e^2] = (\mu_0 V/4\pi)\mathbf{J}_e(\tau_r)$; $v_A \cong \langle(\omega|\Omega_e|/\omega_{pe}^2)^{1/2}\rangle c$ is the average phase velocity of the radiation over the propagation path from the source to the receiver. The wave magnetic induction is obtained from the vector potential by

$$\mathbf{B} = \nabla \times \mathbf{A} = -\left[\frac{1}{r^2} + \frac{1}{r v_A}\frac{\partial}{\partial t}\right](\hat{\mathbf{r}} \times \mathbf{p}) \tag{9.7}$$

On the right-hand side of (9.7), the first term represents the near field and the second term is the radiation (far) field. In the region right beneath the electrojet, the near field is dominant as the modulation frequency $f_1 < 50$ Hz and can be neglected in the VLF regime.

In (9.7), the near field is proportional to $\mathbf{J}_e(\tau_r)$ given by (9.2) and the radiation field is proportional to $\partial\mathbf{J}_e(t - r/v_A)/\partial t$ given by

$$\frac{\partial \mathbf{J}_e}{\partial t} = \left(\frac{5n_0e^2E_0}{6m_e}\right)\left\{\frac{[\hat{\mathbf{x}}(\Omega_e^2 - \nu_{en}^2) - \hat{\mathbf{y}}2|\Omega_e|\nu_{en}]}{(\nu_{en}^2 + \Omega_e^2)^2}\right\}\left(\frac{\nu_{en}}{T_e}\right)\frac{\partial T_e}{\partial t} \tag{9.8}$$

With the aid of (9.2) and (9.8) and the condition $\nu_{en} \ll |\Omega_e|$, (9.7) becomes

$$\begin{aligned}
\mathbf{B} = &-V\left(\frac{5}{24\pi c^2}\right)\left(\frac{E_0}{r}\right)\left[\frac{\omega_{pe}^2}{(\nu_{en}^2 + \Omega_e^2)}\right] \\
&\times\left\{(\hat{\mathbf{r}}\times\hat{\mathbf{x}})\nu_{en}\left[\frac{6}{5r} + \frac{(\Omega_e^2 - \nu_{en}^2)}{(\nu_{en}^2 + \Omega_e^2)}\frac{1}{v_A}\left(\frac{\partial}{\partial t}\ln T_e\right)\right]\right. \\
&\left.-(\hat{\mathbf{r}}\times\hat{\mathbf{y}})|\Omega_e|\left[\frac{6}{5r} - \frac{2\nu_{en}^2}{(\nu_{en}^2 + \Omega_e^2)}\frac{1}{v_A}\left(\frac{\partial}{\partial t}\ln T_e\right)\right]\right\} \\
\cong &-\hat{G}\left(\frac{E_0}{cr}\right)\left\{(\hat{\mathbf{r}}\times\hat{\mathbf{x}})\nu_{en}\left[\frac{6}{5r} + \frac{1}{v_A}\left(\frac{\partial}{\partial t}\ln T_e\right)\right]\right. \\
&\left.-(\hat{\mathbf{r}}\times\hat{\mathbf{y}})\Omega_e\left[\frac{6}{5r} - 2\left(\frac{\nu_{en}}{\Omega_e}\right)^2\frac{1}{v_A}\left(\frac{\partial}{\partial t}\ln T_e\right)\right]\right\}
\end{aligned} \tag{9.9}$$

where $\hat{G} = 5V\omega_{pe}^2/24\pi c\Omega_e^2$. Set $\hat{\mathbf{r}} = \hat{\mathbf{x}}\sin\theta\cos\varphi + \hat{\mathbf{y}}\sin\theta\sin\varphi - \hat{\mathbf{z}}\cos\theta$ in (9.9), where θ and φ are the poloidal and azimuthal angles with respect to the downward direction, the magnitude of $\mathbf{B}$ is given to be

$$\begin{aligned}
\left|\frac{c\mathbf{B}}{E_0}\right| = \hat{G}\frac{1}{r}&\left\{\left[\Omega_e^2\left(\frac{6}{5r}\right)^2 + \nu_{ne}^2\left(\frac{6}{5r} - \frac{1}{v_A}\frac{\partial}{\partial t}\ln T_e\right)^2\right]\cos^2\theta\right. \\
&+\left[(\nu_{en}\sin\varphi + \Omega_e\cos\varphi)\left(\frac{6}{5r}\right) + \frac{\nu_{en}}{\Omega_e}(\Omega_e\sin\varphi - 2\nu_{en}\cos\varphi)\right. \\
&\left.\left.\times\left(\frac{1}{v_A}\frac{\partial}{\partial t}\ln T_e\right)\right]^2\sin^2\theta\right\}^{1/2}
\end{aligned} \tag{9.10}$$

The near field is included in the numerical analyses only for comparison with the ELF experimental results which are measured near the heating site. In the VLF applications, only the radiation field needs to be considered and (9.10) is reduced to

$$\left|\frac{c\mathbf{B}}{E_0}\right| = \hat{G}\frac{1}{r}\nu_{en} \times \left\{\cos^2\theta + \sin^2\theta\left[\sin\varphi - 2\left(\frac{v_{en}}{|\Omega_e|}\right)\cos\varphi\right]^2\right\}^{1/2}\frac{1}{v_A}\frac{\partial}{\partial t}\ln T_e \tag{9.11}$$

9.2.2 *Electron heating*

In the presence of HF heating, the electron temperature is governed by the electron thermal energy equation (1.27). Let T_{eb} be the background electron temperature in the absence of the heating wave and substitute it into (1.27), the solar source power $= \delta(T_{eb})\ \nu_e(T_{eb})(T_{eb} - T_n) - (2/3)(Q_{E0}/n_0)$ is determined, here $Q_{E0} = \nu_{en}(T_{eb})n_0m_eu_e(T_{eb})^2$ is the Ohmic heating power density contributed by the background (unmodulated) electrojet current. The total Ohmic heating power density is given by

$$Q = \left\langle\frac{J_{et}^2}{\sigma_0}\right\rangle \cong \nu_{en}n_0m_e[u_e^2 + \langle|v_{pe}|^2\rangle] = Q_E + Q_H \tag{9.12}$$

where $\mathbf{J}_{et} = -en(\mathbf{u}_e + \mathbf{v}_{pe})$ is the total electron current density in the background plasma carrying an electrojet and interacting with the HF heater, angle brackets indicate the time average over the HF wave period, and $\sigma_0 = n_0e^2/m\nu_{en}$ is the conductivity of the plasma, which causes Ohmic loss of the HF heater; $\mathbf{u}_e$ is given by (9.1) and $\mathbf{v}_{pe}$ is the quiver velocity of electrons imposed by the HF wave electric field; $Q_E = \nu_{en}n_0m_eu_e^2$ and $Q_H = \nu_{en}n_0m_e\langle|v_{pe}|^2\rangle$, where $u_e = eE_0/m_e|\Omega_e|$ and $\langle|v_{pe}|^2\rangle = v_q^2M(t)$ depends on the magnitude of the electron quiver speed v_q and the modulation form $M(t)$ of the HF heating power. An intensity-modulated HF heater can be radiated directly by the HF transmitter. It can also be set up through a beat wave approach by overlapping two CW HF heaters in the electrojet, where the two HF heaters have a frequency difference in the VLF band.

9.2.3 *Electron heat loss in collisions*

The fractional energy loss term on the LHS of the electron thermal energy equation (1.27) involves both elastic and inelastic collisions. The main processes involved in the inelastic collisions in the energy range of interest (<6 eV) are the rotational and vibration excitation of N_2 and O_2. Loss through optical excitation is neglected. The fractional electron heat loss rate, $\delta(T_e)\nu_e(T_e) = (\delta_{el} + \delta_r + \delta_v)\nu_e$, is determined to be

$$\delta(T_e)\nu_e(T_e) \cong 2\frac{m_e}{m_n}\nu_{en0}[\chi^{5/6} + 8.38\chi^{-1/2} + \mu_I(\chi)] \qquad (9.13)$$

where a normalized temperature $\chi = T_e/T_{e0}$ is introduced, $\nu_{en0} = \nu_{en}(T_e = T_{e0})$ and a reference equilibrium electron temperature $T_{e0} = 1500\,K$ is assumed; and

$$\begin{aligned}\mu_I(\chi) = \chi^{-5/2}\{&14.73\exp[6.98\mathfrak{R}(\chi)] + 0.1\exp[13.88\mathfrak{R}(\chi)] \\ &+ 3.81\times10^{-4}\exp[19.34\mathfrak{R}(\chi)] + 1.43\times10^{-4}\exp[23.67\mathfrak{R}(\chi)] \\ &+ 2.7\times10^{-5}\exp[26.1\mathfrak{R}(\chi)] + 2.59\times10^{-6}\exp[28.05\mathfrak{R}(\chi)] \\ &+ 3.54\times10^{-7}\exp[30.84\mathfrak{R}(\chi)]\}\end{aligned} \qquad (9.14)$$

where $\mathfrak{R}(\chi) = 1 - 1/\chi$.

The RHS of (9.13) is a sum of two main causes from (1) elastic collisions with the neutral particles ($\propto \chi^{5/6}$) and from (2) inelastic collisions associated with rotational ($\propto \chi^{-1/2}$) and vibration ($\propto \chi^{-5/2}$) excitations of the neutral particles.

9.2.4 *Numerical model of very- low-frequency wave radiated by heater-modulated electrojet*

In the numerical analysis, dimensionless variables and parameters are introduced:

$$\nu_{en}/\nu_{en0} = \chi^{5/6}, \quad \tau = \nu_{en0}t, \quad \eta = (\nu_{en0}/v_{t0})z,$$
$$R = (\nu_{en0}/v_A)r, \quad \xi = \tau - R,$$
$$|c\mathbf{B}(\mathbf{r},t)/E_0| = B(\xi), \quad \omega_{10} = \omega_1/\nu_{en0}, \quad T_{e0} = 1500\,\mathrm{k} \text{ is set},$$
$$v_{t0} = (T_{e0}/m_e)^{1/2} = 1.5\times 10^5\,\mathrm{m/s}, \quad \eta = (eE_0/m_e\Omega_e v_{t0})^2,$$
$$b = (\nu_{en0}/\Omega_e)^2, \quad \text{and} \quad \alpha = (v_q/v_{t0})^2;$$

and set $\theta = 0$ in (9.11) to simplify the analyses.

Thus, (9.12) and (9.11) are normalized to the dimensionless forms

$$\begin{aligned}\frac{d\chi}{d\tau} &+ 2\frac{m_e}{m_n}\{[\chi^{5/6} + 8.38\chi^{-1/2} + \mu_I(\chi)](\chi - \zeta) - \Delta\}\\ &= \frac{2\alpha m_n}{3}\chi^{5/6} + \frac{2\eta}{3}\left[\frac{\chi^{5/6}}{(1+b\chi^{5/3})} - \frac{\chi_b^{5/6}}{(1+b\chi_b^{5/6})}\right]\\ &+ 2\left[\frac{\chi^{1/4}}{n_0(\chi)}\right]\frac{d}{d\eta}\left\{[n_0(\chi)\chi^{-1/12}]\frac{d\chi}{d\eta}\right\}\end{aligned} \tag{9.15}$$

and

$$B(\xi) = \Lambda_0\left\{\frac{\chi^{-1/6}(\xi)}{[1+b\chi^{5/3}(\xi)]}\right\}\frac{d\chi(\xi)}{d\xi} \tag{9.16}$$

where $\zeta = T_n/T_{e0}, \Delta = [\chi_b^{5/6} + 8.38\chi_b^{-1/2} + \mu_I(\chi_b)](\chi_b - \zeta)$, and $\chi_b = T_{eb}/T_{e0}$; $\Lambda_0 = (5\,V\omega_{pe}^2\Omega_e b^{3/2}/24\pi c v_A^2 R)$.

Only temporal modulation is considered, the spatial derivative term on the RHS of (9.15) is set to zero which reduces (9.16) to

$$\frac{d\chi}{d\tau} + 2\frac{m_e}{m_n}\{[\chi^{5/6} + 8.38\chi^{-1/2} + \mu_I(\chi)](\chi - \zeta) - \Delta\}$$
$$= \frac{2\alpha m_n}{3}\chi^{5/6} + \frac{2\eta}{3}\left[\frac{\chi^{5/6}}{(1 + b\chi^{5/3})} - \frac{\chi_b^{5/6}}{(1 + b\chi_b^{5/6})}\right] \quad (9.17)$$

9.2.5 *Numerical results*

Three cases of modulation frequency $f_1 = 1, 3$, and 10 kHz are presented as examples. The numerical analysis is carried out for electrojet appearing near 90 km, where $T_{eb} = 200$ K, $T_n \cong T_i \cong 190$ K, and $E_0 = 50$ mV/m. A heater frequency of $\omega_0/2\pi = 3.2$ MHz is used; $\nu_{en0} = 9 \times 10^5 s^{-1}, m_n/m_e = 5.52 \times 10^4, |\Omega_e|/2\pi = 1.4$ MHz $\gg \nu_{en0}$ and $v_q = 1.56 \times 10^4 \varepsilon_{p0}$ m/s, where ε_{p0} is in V/m, and c is the speed of light in free space. Assume $\varepsilon_{p0} = 1.5$ V/m, which leads to $v_q = 2.33 \times 10^4$ m/s, $\alpha = 0.024$, and consider rectangular wave modulation having 50% duty cycle, i.e., $M_R(\tau) = [1/2 + \sum_{k=1} \sin c(k\pi/2) \cos k(\omega_{10}\tau + \pi/2)]$.

Equation (9.17) is solved numerically subject to the initial condition $\chi(0) = 1$. Time functions of modulated electron temperature produced by the X-mode heater at the three modulation frequencies are obtained as shown in Fig. 9.1(a). The results are substituted into (9.16) to obtain the radiation fields $B(\xi)$ in the three cases, which are presented in Fig. 9.1(b), in which the vertical axis has an arbitrary unit. As shown, in the cases of 1 and 3 kHz modulations, the on periods are long enough for the heating to reach a steady state level; but the off period in the 3 kHz modulation is too short for cooling, thus the temperature modulation amplitude drops. Both heating and cooling times are too short in the case of 10 kHz modulation; it takes the modulation three periods to reach a steady state, which settles at a much lower modulation amplitude. Consequently, the radiation intensity decreases with the increase of the modulation frequency as demonstrated in Fig. 9.1(b), in which each signal also contains harmonic components. Taking a Fourier transform, the time harmonic component at the modulation frequency (i.e., the first harmonic component) is extracted; the dependency of the amplitude on the modulation frequency is presented in Fig. 9.1(c). As shown, the amplitude of the VLF signal, generated via electrojet modulation, decreases monotonically with the frequency.

The heating and cooling rates as well as the available heating and cooling times in one modulation period affect the efficacy of conductivity modulation by an intensity-modulated HF heater. Consequently, the efficacy of VLF generation via electrojet modulation is expected to decrease as the modulation frequency increases.

9.3 Very- Low-Frequency Wave Generation by Beating of Two High-Frequency Heater Waves

The momentum equations (1.26) of electrons and ions are combined; in the resultant equation, those terms contributing directly to the nonlinear beating are equated to be

$$\mathrm{J} \times \hat{\mathbf{z}} = -\left(\frac{\mathrm{e}}{|\Omega_\mathrm{e}|}\right)\left[\boldsymbol{\nabla} \cdot (\mathrm{n v_e v_e}) + \frac{1}{\mathrm{m_e}}\boldsymbol{\nabla}\mathrm{P_e}\right] \tag{9.18}$$

where $\mathbf{J} = -\mathrm{ne}(\mathbf{v}_\mathrm{e} - \mathbf{v}_\mathrm{i})$; the electron and ion inertial terms and the ion convective and pressure terms, which do not directly contribute to the nonlinear beating, are neglected; the relations $\mathrm{m_e n_e \nu_{ei} = m_i n_i \nu_{ie}}$ and $\mathrm{n_e \cong n_i = n}$, and the conditions $\nu_\mathrm{en} \ll |\Omega_\mathrm{e}|$ and $\nu_\mathrm{in} \ll \Omega_\mathrm{i}$, are applied to simplify the equation. The two terms on the RHS of (9.18) correspond to the ponderomotive and thermal pressure forces.

9.3.1 *Beating current*

Equation (9.18) is then solved to obtain the nonlinear beating current density $\mathbf{J}_\mathrm{B}$, at the difference frequency of the two HF heater waves, to be

$$\mathbf{J}_\mathrm{B} = -\left(\frac{\mathrm{e}}{|\Omega_\mathrm{e}|}\right)\hat{\mathbf{z}} \times \left\langle \boldsymbol{\nabla} \cdot (\mathrm{n_e}\mathbf{V}_\mathrm{pe}\mathbf{V}_\mathrm{pe}) + \frac{1}{\mathrm{m_e}}\boldsymbol{\nabla}\mathrm{P_e}\right\rangle = \mathbf{J}_\mathrm{BP} + \mathbf{J}_\mathrm{BT} \tag{9.19}$$

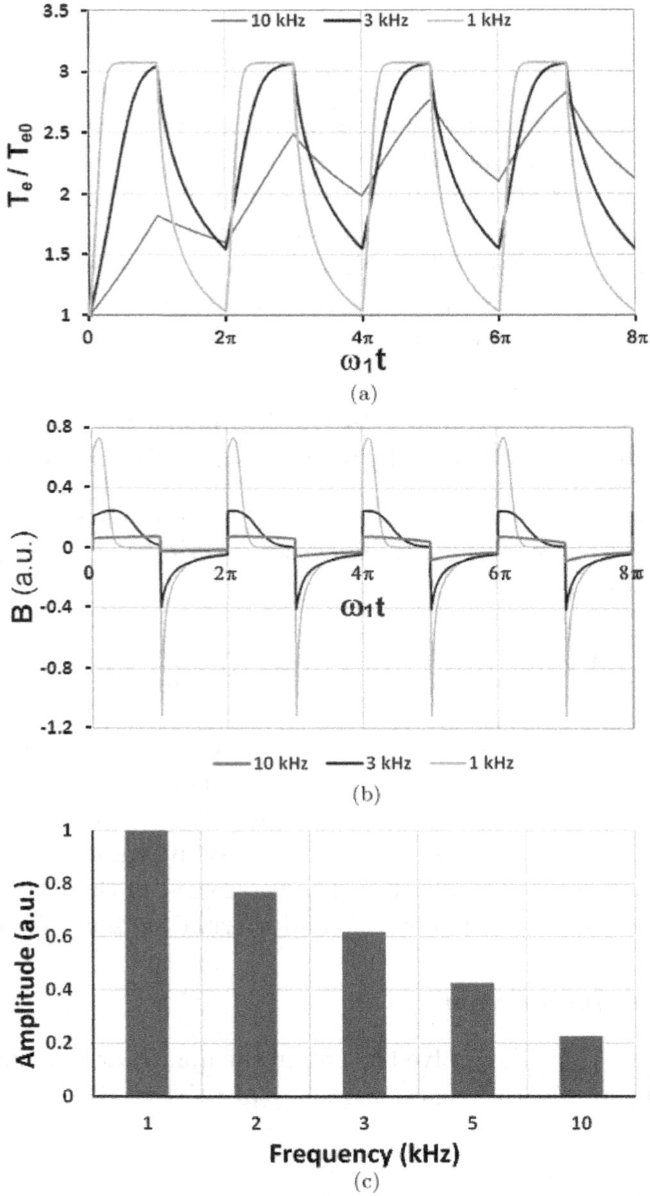
10 kHz
3 kHz
1 kHz
T_e / T_{e0}
$\omega_1 t$
0
2π
4π
6π
8π
3.5
3
2.5
2
1.5
1
(a)
B (a.u.)
0.8
0.4
0
-0.4
-0.8
-1.2
(b)
Amplitude (a.u.)
Frequency (kHz)
1
2
3
5
10
0.8
0.6
0.4
0.2
(c)

Fig. 9.1. (*figure on facing page*) The intensity of X-mode heater is modulated by a rectangular wave at 1 to 10 kHz; (a) electron temperature modulation and (b) the magnetic induction of the radiation of the modulated electrojet in the cases of 1, 3, and 10 kHz modulations; and (c) the relative amplitudes of the radiations at 1, 2, 3, 5, and 10 kHz.

where $\mathbf{J}_{BP}$ and $\mathbf{J}_{BT}$ are driven by the ponderomotive and thermal pressure forces, respectively; $\mathbf{V}_{pe}$ is the electron quiver velocity induced by the HF fields, and the electron density $n_e = n_0 + \Delta n_e$; sum of the background electron density n_0 and the density Δn_e of irregularities; $\langle\ \rangle$ represents a VLF bandpass filter; $P_e = nT_e$ and T_e is governed by (1.27).

Only temporally modulated heating is considered, the convective and diffusion terms are set to zero in (1.27), which becomes

$$\frac{\partial T_e}{\partial t} + \delta(T_e)\nu_e(T_e)(T_e - T_n) + 2\nu_{ei}\frac{m_e}{m_i}(T_e - T_i)$$
$$= \frac{2}{3}\nu_{el} m_e \langle|\mathbf{V}_{pe}|^2\rangle + \delta(T_{e0})\nu_e(T_{e0})(T_{e0} - T_n) \qquad (9.20)$$

where $\delta(T_e)\nu_e(T_e)$ is the average fraction of energy loss rate in (elastic and inelastic) collisions with the neutrals.

Let $T_e = T_{e0} + \langle\delta T_e\rangle + \cdots$ and $\langle|\mathbf{V}_{pe}|^2\rangle = W\exp(-i\omega_1 t) + \text{c.c.}$, the induced temperature perturbation $\langle\delta T_e\rangle$, which oscillates at the beat frequency ω_1, is obtained to be

$$\langle\delta T_e\rangle = i\frac{2}{3}m_e\left[\frac{\nu_{el}}{(\omega_1 + i\delta\nu)}\right]W\exp(-i\omega_1 t) + \text{c.c.} \qquad (9.21)$$

where $\nu_{el} = \nu_{ei} + \nu_{en}$ is the elastic collision frequency and $\delta\nu = \delta(T_{e0})$ $\nu_e(T_{e0}) + 2\nu_{ei}(m_e/m_i)$.

9.3.2 *Operation of the high-frequency transmitter array for beat wave generation*

The HF transmitter array is split into two sub-arrays, transmitting CW heaters at slightly different frequencies f_{01} and f_{02} along the geomagnetic zenith as illustrated in Fig. 9.2, where $f_{01} = f_0, f_{02} = f_0 + f_1$, and f_1 is the beat wave frequency in the VLF band.

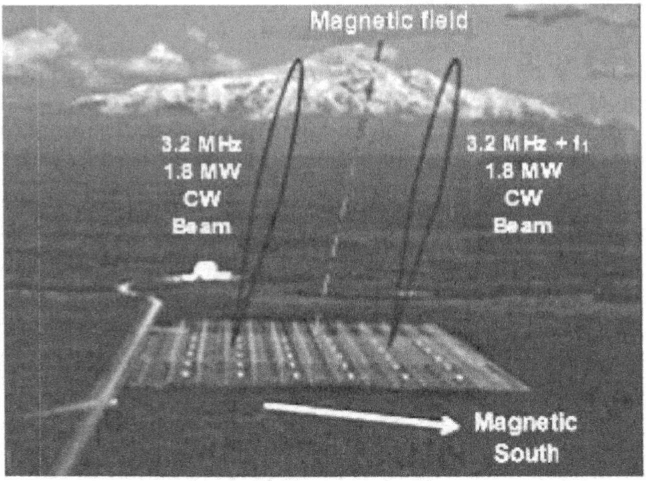

Fig. 9.2. The HAARP HF transmitter facility at Gakona, Alaska, operates in two sub-antenna array arrangement for beat wave generation. It is transmitted at full power (3.6 MW) with HF heater of 3.2 MHz directed along the geomagnetic zenith.

The wave field $\mathbf{E}_p$ of overlapped heaters is

$$\begin{aligned}\mathbf{E}_p &= \mathbf{E}_{p1} + \mathbf{E}_{p2}\\ &= \frac{1}{2}(\hat{\mathbf{x}} \pm i\hat{\mathbf{y}})E_{p0}\{1 + \exp[-i(\omega_1 t - \Psi)]\}\exp[i(k_0 z - \omega_0 t)] + \text{c.c.}\end{aligned} \tag{9.22}$$

where "±" correspond to O/X-mode heaters; E_{p0} is the field amplitude of each heater beam, ψ is the phase difference of two HF waves transmitted by the two sub-arrays; ω_0 and k_0 are the heater radian frequency and wavenumber.

The beat wave field (9.22) is plotted in Fig. 9.3, where the carrier is plotted with a (much enlarged) different time scale. As shown the beat wave has a temporally modulated intensity. The induced electron velocity is derived to be

$$\begin{aligned}\mathbf{V}_{pe} &= -\frac{1}{2}i(\hat{\mathbf{x}} \pm i\hat{\mathbf{y}})\left[\frac{eE_{p0}}{m_e(\omega_0 \pm \Omega_e)}\right]\\ &\quad\times\{1 + \exp[-i(\omega_1 t - \Psi)]\}\exp[i(k_0 z - \omega_0 t)] + \text{c.c.}\end{aligned} \tag{9.23}$$

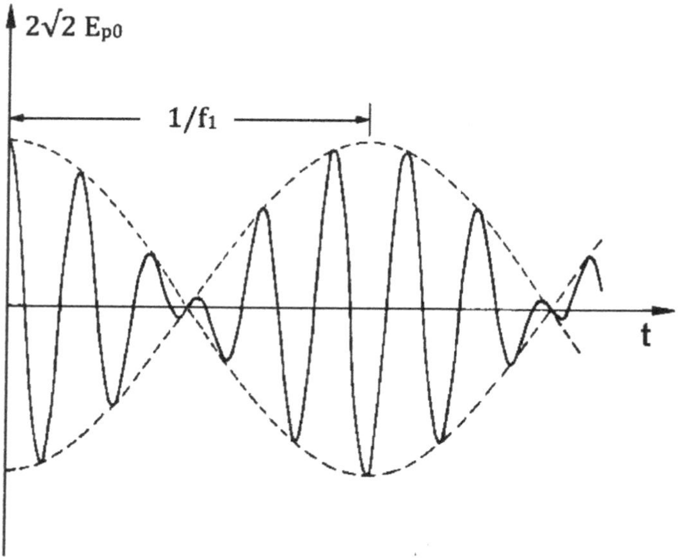

Fig. 9.3. Beat wave.

Thus, $W = [eE_{p0}/m(\omega_0 \pm |\Omega_e|)]^2 e^{i\psi}$. With the aid of (9.21) and (9.23), (9.19) becomes

$$\begin{aligned}
\mathbf{J}_B(\mathbf{r}, t) &= \left(\frac{e}{|\Omega_e|}\right)\left[\frac{e}{m_e(\omega_0 \pm |\Omega_e|)}\right]^2 \\
&\quad \times \left\{\left(\hat{\mathbf{y}}\frac{\partial}{\partial x} + \hat{\mathbf{x}}\frac{\partial}{\partial y}\right)(n_e E_{p0}^2)\cos(\omega_1 t - \Psi)\right. \\
&\quad - \frac{4}{3}\left[\frac{\nu_{el}\omega_1}{(\omega_1^2 + \delta\nu^2)}\right]\left(\hat{\mathbf{y}}\frac{\partial}{\partial x} - \hat{\mathbf{x}}\frac{\partial}{\partial y}\right)(n_e E_{p0}^2) \\
&\quad \left. \times\ [\sin(\omega_1 t - \Psi) + (\delta\nu/\omega_1)\cos(\omega_1 t - \Psi)]\right\} \\
&= [\mathbf{J}_{BP}(\mathbf{r}) + J_{BT}(\mathbf{r})]\exp(-i\omega_1 t) + \text{c.c.} \qquad (9.24)
\end{aligned}$$

where $\mathbf{J}_{BP}(\mathbf{r}) = (e/2|\Omega_e|)(\hat{\mathbf{y}}\partial/\partial x + \hat{\mathbf{x}}\partial/\partial y)(n_e W)$ and $\mathbf{J}_{BT}(\mathbf{r}) = -i(2e/3|\Omega_e|)[1 - i(\delta\nu/\omega_1)][\nu_{el}\omega_1/(\omega_1^2 + \delta\nu^2)](\hat{\mathbf{y}}\partial/\partial x - \hat{\mathbf{x}}\partial/\partial y)(n_e W)$; the $(\omega_0 \pm |\Omega_e|)^{-2}$ dependence of W indicates that X-mode heaters are more effective than O-mode heaters to generate the beat wave current.

9.3.3 *Radiation field of the beat wave*

The magnetic flux density of the radiation from the time harmonic current source (9.24) is given by

$$\mathbf{B}(\mathbf{r}, t) = \nabla \times \mathbf{A}(\mathbf{r}) \exp(-i\omega_1 t) + \text{c.c.} \tag{9.25}$$

where the vector potential $\mathbf{A}(\mathbf{r})$ is

$$\mathbf{A}(\mathbf{r}) = \left(\frac{\mu_0 e^{ikr}}{r}\right) \int [\mathbf{J}_{BP}(\mathbf{r}') + J_{BT}(\mathbf{r}')] dV' \tag{9.26}$$

k is the wavenumber of the radiation and $dV' = dx'dy'dz'$ is the differential volume of the induced current distribution at $\mathbf{r} = \mathbf{r}'$.

As shown, an electrojet-independent ionospheric VLF transmitter can be set up by applying two overlapped HF heater waves of slightly different frequencies to the ionosphere. Although the beating may be set at any region of the ionosphere, it is the most effective in the F region, where the cross-field electron current may be produced effectively. The radiation depends on the HF heater intensity, rather than the electrojet strength. Thus the beat wave approach generates VLF radiations over a larger frequency band than by the modulated electrojet. These combined characteristics render the beat wave approach to be a potentially reliable, broadband ionospheric VLF transmitter.

9.4 Experiments and Results of Beat Wave Generation

HF heating experiments were performed at HAARP on July 26 and July 27, 2011 to explore beat wave generation of VLF waves. The 12×15 array of the HAARP transmitter was split into two 6×15 sub-arrays run at CW full power (1.8 MW each) to transmit X-mode HF waves along the geomagnetic zenith; one sub-array transmitted at $f_0 = 2.85$ MHz $\sim 2f_{ce}$, twice the electron cyclotron frequency,

f_{ce}, and the other at $f_0 + f_1$, where $f_1 = 3.5$, 5.5, 7.6, and 9.5 kHz. This beat wave scheme for VLF wave generation run at a single beat frequency for the 60 second period. The experiment was performed (1) from 09:42:00 to 09:50:30 and (2) from 10:12:00 to 10:20:30 on July 26, and (3) from 10:14:00 to 10:22:30 on July 27.

The foF2s of the ionosphere in the three experimental periods are around 2.55, 1.8, and 3.55 MHz, respectively. The x-mode heating wave of 2.85 MHz is cutoff at $f_{pc} = [f_0(f_0 - f_{ce})]^{1/2} \sim 2.015$ MHz; thus, the ionosphere was overdense in the first experimental period, underdense in the second experimental period, and well overdense in the third experimental period. Under the three distinctive ionospheric situations, the HF heaters were (1) reflected slightly below, and (2) penetrating through, and (3) reflected well below, the F-peak (i.e., foF2 layer).

The VLF receiver recorded both North–South (N–S) and East–West (E–W) components of the wave magnetic field. The N–S component dominated in all measurements. The power spectra of the radiations, represented by the N–S component, are presented in Fig. 9.4. These results show that the VLF emissions generated in the first experimental period, presented in Fig. 9.4(a), are much stronger than those, presented in Figs. 9.4(b) and 9.4(c), generated in the other two periods, respectively. The VLF emissions generated in the second experimental period, when the ionosphere was underdense to the HF heaters, have the lowest spectral intensities. The spectral intensities that appeared in the third experimental period when the ionosphere was well overdense are stronger than those that appeared in the second experimental period and have an increasing dependence on the frequency. In the first experimental period, the ionosphere was also overdense, but the HF reflection height was close to the foF2 layer. The spectral intensities of the VLF emissions are further increased considerably. Moreover, an anomaly at 5.5 kHz is shown, where the spectral intensity is about 7 dB above the normal, deduced from the intensity trend with frequency in the other two experimental periods. 5.5 kHz matches the lower hybrid resonance frequency in the lower F region, where $\omega_{pe} = 0.883|\Omega_e|$. Thus, lower hybrid waves could be excited by the HF heaters in

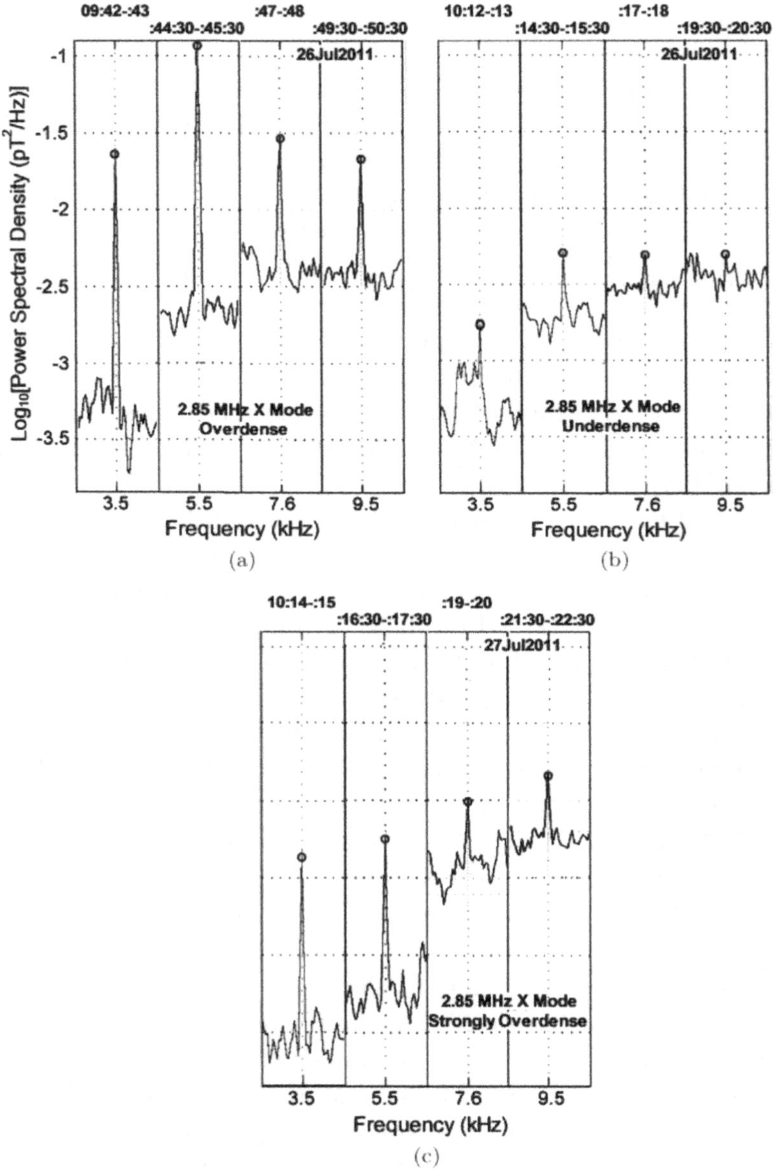

Fig. 9.4. The spectral power densities of the VLF radiation at 4 different frequencies (3.5, 5.5, 7.6, and 9.5 kHz), generated (a) from 09:42:00 to 09:50:30 on July 26 (when the ionosphere was overdense), (b) from 10:12:00 to 10:20:30 on July 26 (underdense), and (c) from 10:14:00 to 10:22:30 on July 27 (strongly overdense).

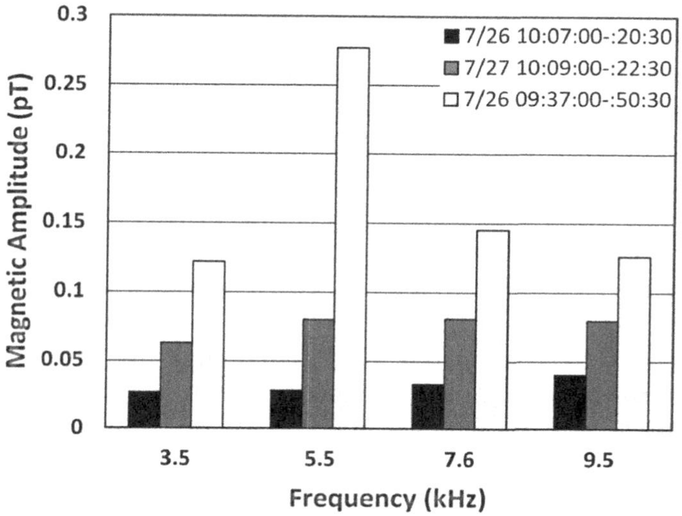

Fig. 9.5. Comparison of the magnetic field intensities of the VLF radiations in three experiments.

the case of $f_1 = 5.5$ kHz and the excited lower hybrid waves were then scattered by the background density irregularities, similar to the stimulated electromagnetic emission (SEE) process, to generate 5.5 kHz radiation; it may explain the 7 dB anomalous enhancement of 5.5 kHz radiation intensity in the first experimental period, during which the ionograms indeed indicate the presence of intense large-scale density irregularities in the background ionosphere.

The power spectra are integrated over the frequency to obtain the field amplitudes, which are presented in Fig. 9.5 for comparison at each emission frequency. Again, an anomaly at 5.5 kHz is clearly seen.

Although the beat wave mechanism for VLF generation depends strongly on the background conditions, it does not rely on the presence of electrojet; moreover, the experimental results show that VLF wave is generated with good signal to noise (S/N) ratio in three distinctly different background conditions. It is also relatively broadband. These results support that this electrojet-independent mechanism is viable for establishing a reliable ionospheric virtual VLF transmitter.

Problems

P.9.1. Show that the electron quiver velocity $\mathbf{v}_{pe}$ in a circularly polarized HF field $\mathbf{E}_p = \boldsymbol{\varepsilon}_p + \text{c.c.}$ is given by $\mathbf{V}_{pe} = -i[e/m(\omega_0^2-\Omega_e^2)](\omega_0\overleftrightarrow{\mathbf{I}}\cdot - i\Omega_e\hat{\mathbf{z}}\times)\boldsymbol{\varepsilon}_p + \text{c.c.}$, where ω_0 is the carrier frequency of the HF heaters.

P.9.2. Use the formula derived in P9.1 to verify (9.23) if $\mathbf{E}_p$ is given by (9.22).

P.9.3. The radiation resistance R_r of a linear antenna is defined to be $R_r = 2P_r/I_m^2$, where P_r is the time-averaged radiation power and I_m is the maximum amplitude of the antenna current. Drive a time harmonic current $I(z,t) = I_m \sin(kL/2 - k|z|)\cos\omega t$ in a linear dipole antenna of length L, where $k = \omega/c = 2\pi/\lambda$, show that

$$R_r = 60\int_0^\pi d\theta \left[\cos\left(\frac{1}{2}kL\cos\theta\right) - \cos\left(\frac{1}{2}kL\right)\right]^2 \Big/ \sin\theta$$

P.9.4. Show that $R_r = 20(\pi L/\lambda)^4$ for $kL \ll 1$.

Bibliography

Abramowitz, M. and I. A. Stegun, *Handbook of Mathematical Functions with Formulas, Graphs and Mathematical Tables*, National Bureau of Standards, Washington, D. C., 1964.

Alfvén, H., "On the existence of electromagnetic–hydrodynamic waves", *Arkiv. Mat. Astron. Fysik.*, **29B**(2), 1942.

Braginski, S. I., Transport Processes in a Plasma, in *Review of Plasma Physics*, M. A. Leontovich, (ed.). Consultants Bureau, New York, Vol. 1, pp. 205–311, 1965.

Budden, K. G., *Radio Waves in the Ionosphere*, Cambridge University Press, Cambridge, 1961.

Chen, F., *Introduction to Plasma Physics*, Plenum, New York, 1974.

Davidson, R. C., *Methods in Nonlinear Plasma Theory*, Academic Press, New York, 1972.

Fejer, J. A., "Ionospheric modification and parametric instabilities", *Rev. Geophys.*, **17**, 135–153, 1979.

Frey, A., P. Stubbe and H. Kopka, "First experimental evidence of HF produced electron density irregularities in the polar ionosphere; disgnosed by UHF radio star scintillations", *Geophys. Res. Lett.*, **11**, 523, 1984.

Fried, B. D. and S. D. Conte, *The Plasma Dispersion Function*, Academic Press, New York, 1961.

Gardner, S., J. M. Greene, M. D. Kruskal, and R. M. Miura, "Method for solving the Kortewe–deVries equation", *Phys. Rev. Lett.*, **19**(19), 1095–1097.

Ginzburg, V. L., *The Propagation of Electromagnetic Waves in Plasmas*, 2nd edn., Pergamon Press, New York, 1970.

Gordon, W. E. and H. C. Carlson, "Arecibo heating experiments", *Radio Sci.*, **9**, 1041, 1974.

Gurevich, V. A., *Nonlinear Phenomena in the Ionosphere*, Springer-Verlag, New York, 1978.

Hagfors, T., W. Kofman, H. Kopka, P. Stubbe, and T. Aijanen, "Observations of enhanced plasma lines by EISCAT during heating experiments", *Radio Sci.*, **18**, 861, 1983.

Hayes, N., "Damping of plasma oscillation in the linear theory", *Phys. Fluids*, **4**, 1387, 1961.

Huang, J. and S. P. Kuo, "A theoretical model for the broad up-shifted maximum in the stimulated electromagnetic emission spectrum", *J. Geophys. Res.*, **99**(A10), 19569–19576, 1994.

Huang, J. and S. P. Kuo, "A generation mechanism for the downshifted peak in stimulated electromagnetic emission spectrum", *J. Geophys. Res.*, **100**, 21433–21438, 1995.

Ichimaru, S., *Basic Principles of Plasma Physics: A Statistical Approach*, Benjamin Frontiers in Physics, 1973.

Kaw, P. K., W. L. Kruer, C. S. Liu, and K. Nishikawa, *Advances in Plasma Physics*, A. Simon and W. B. Thompson, (eds.). John Wiley & Sons, New York, Vol. 6, 1976.

Kossey, P., J. Heckscher, H. Carlson, and E. Kennedy, "HAARP: High frequency active auroral research program", *J. Arctic Res. U. S.*, **1**, 1, 1999.

Krall, N. A. and A. W. Trivelpiece, *Principle of Plasma Physics*, McGraw-Hill, New York, 1973.

Kuo, S. P., B. R. Cheo, and M. C. Lee, "The role of parametric decay instabilities in generating ionospheric irregularities", *J. Geophys. Res.*, **88**, 417, 1983.

Kuo, S. P. and G. Schmidt, "Filamentation instability in magneto plasmas", *Phys. Fluids*, **26**, 2529–2536, 1983.

Kuo, S. P. and M. C. Lee, "Earth magnetic field fluctuations produced by filamentation instabilities of electromagnetic heater waves", *Geophys. Res. Lett.*, **10**(10), 979–981, 1983.

Kuo, S. P. and F. T. Djuth, "A thermal instability for the spread F echoes from the HF-heated ionosphere", *Geophys. Res. Lett.*, **15**(12), 1345–1348, 1988.

Kuo, S. P., M. C. Lee, and J. A. Fejer, "The role of nonlinear beating currents in the theory of parametric instabilities", *J. Geophys. Res.*, **98**(A6), 9515–9518, 1993.

Kuo, S. P., "The role of nonlinear beating currents on parametric instabilities in magnetoplasmas", *Phys. Plasmas*, **3**(11), 3957–3965, 1996.

Kuo, S. P. and M. C. Lee, "Cascade spectrum of HFPLs generated in HF heating experiments", *J. Geophys. Res.*, **110**, A01309, 2005.

Kuo, S. P., Steven S. Kuo, James T. Huynh, and Paul Kossey, "Precipitation of trapped relativistic electrons by amplified whistler waves in the Magnetosphere", *Phys. Plasmas*, **14**(6), 009706 (1–7), 2007.

Kuo, S. P. and A. Snyder, "Observation of artificial Spread-F and large region ionization enhancement in an HF heating experiment at HAARP", *Geophys. Res. Lett.*, **37**, L07101, 2010.

Kuo, S. P., Arnold Snyder, and Chia-Lie Chang, "Electrojet-independent ionospheric ELF/VLF wave generation by powerful HF waves", *Phys. Plasmas*, **17**(8), 082904, 2010.

Kuo, S. P., "Electron cyclotron harmonic resonances in high-frequency heating of the ionosphere", *Phys. Plasmas*, **20**, 092124, 2013.

Kuo, S. P. and A. Snyder, "Artificial plasma cusp generated by upper hybrid instabilities in HF heating experiments at HAARP", *J. Geophys. Res.*, **118**(5), 2734–2743, 2013.

Kuo, S. P., W. T. Cheng, R. Pradipta, M. C. Lee, and A. Snyder, "Observation and theory of whistler wave generation by high-power HF waves", *J. Geophys. Res.*, **118**(3), 1331–1338, 2013.

Kuo, S. P., A. Snyder, and M. C. Lee, "Experiments and theory on parametric instabilities excited in HF heating experiments at HAARP", *Phys. Plasmas*, **21**(7), 062902-1 to 062902-10, 2014.

Kuo, S. P., "On the nonlinear plasma waves in the high-frequency (HF) wave heating of the ionosphere", *IEEE Trans. Plasma Sci.*, **42**(4), 1000–1005, 2014.

Kuo, S. P., "Ionospheric modifications in high frequency heating experiments", *Phys. Plasmas*, **22**(1), 012901 (1–16), 2015.

Kuo, S. P., "Ionospheric very low frequency transmitter", *Phys. Plasmas*, **22**(2), L022901 (1–10), 2015.

Kuo, S. P., "Nonlinear upper hybrid waves and the induced density irregularities", *Phys. Plasmas*, **22**(9), 082904, 2015.

Kuo, S. P. and M. C. Lee, "On the VLF wave generation by beating of two HF heaters", *Phys. Plasmas*, **24**(2), 022902, 2017.

Landau, L. D., "On the Vibrations of the Electronic Plasma", *J. Phys.* (U.S.S.R.), **10**, 25, 1946.

Mishin, E. and T. Pedersen, "Ionizing wave via high-power HF acceleration", *Geophys. Res. Lett.*, **38**, L01105, 2011.

Montgomery, D. C. and D. A. Tidman, *Plasma Kinetic Theory*, McGraw-Hill, New York, 1964.

Nicholson, D. R., *Introduction to Plasma Theory*, John Wiley & Sons, New York, 1983.

Papadopoulos, K. and C. L. Chang, "Generation of ELF/ULF waves in the ionosphere by dynamo processes", *Geophys. Res. Lett.*, **12**, 279, 1985.

Pedersen, T., B. Gustavsson, E. Mishin, E. MacKenzie, H. C. Carlson, M. Starks, and T. Mills, "Optical ring formation and ionization production in high power HF heating experiments at HAARP", *Geophys. Res. Lett.*, **36**, 18, 2009.

Pedersen, T., B. Gustavsson, E. Mishin, E. Kendall, T. Mills, H. C. Carlson, and A. L. Snyder, "Creation of artificial ionospheric layers using high-power HF waves", *Geophys. Res. Lett.*, **37**, L02106, 2010.

Perkins, F. W., C. Oberman, and E. J. Valeo, "Parametric instabilities and ionospheric modification", *J. Geophys. Res.*, **79**, 1478–1496, 1974.

Poeverlein, H., "Strahlenwege von Radiowellen in der Ionosphäre (Ray paths of radio waves in the ionosphere)", *Sitzungsber. Bayer. Akad. Wiss. Math. Naturwiss. Kl.*, **78**, 175–178, 1948.

Reinisch, B. W., I. A. Galkin, G. M. Khmyrov, A. V. Kozlov, K. Bibl, I. A. Lisysyan, G. P. Cheney, X. Huang, D. F. Kitrosser, V. V. Paznukhov, Y.

Luo, W. Jones, S. Stelmash, R. Hamel, and J. Grochmal, "New digisonde for research and monitoring applications", *Radio Sci.*, **44**, RS0A24, 2009.

Sagdeev, R. Z. and A. A. Galeev, *Nonlinear Plasma Theory*, T. M. O'Neil and D. L. Book, (eds.). Benjamin Reading, Massachusetts, 1969.

Schmidt, G., *Physics of High Temperature Plasmas*, 2nd edn., Academic Press, New York, 1979.

Spitzer, L. J. Jr., *Physics of Fully Ionized Gases*, Wiley (Interscience), New York, 1962.

Stix, T. H., *The Theory of Plasma Waves*, McGraw-Hill, New York, 1962.

Stix, T. H., *Waves in Plasmas*, Springer-Verlag, New York, 1992.

Stubbe, P., H. Kopka, and R. L. Dowen, "Generation of ELF and VLF waves by polar electrojet modulation: Experimental results", *J. Geophys. Res.*, **86**, 9073, 1981.

Stubbe, P., H. Kopka, B. Thide, and H. Derblom, "Stimulated electromagnetic emission: A new technique to study the parametric decay instability in the ionosphere", *J. Geophys. Res.*, **89**, 7523–7536, 1984.

Stubbe, P., H. Kopka, M. T. Rietveld, A. Frey, P. Hoeg, H. Kohl, E. Nielsen, and G. Rose, "Ionospheric modification experiments with the Tromso heating facility", *J. Atoms. Terr. Phys.*, **47**, 1151, 1985.

Vlasov, A. A., "On the kinetic theory of an assembly of particles with collective interaction", *J. Phys.* (U.S.S.R.), **9**, 25, 1945.

Index